Analog Circuits and Signal Processing

Series Editors:

Mohammed Ismail, Dublin, USA
Mohamad Sawan, Montreal, Canada

The *Analog Circuits and Signal Processing* book series, formerly known as the *Kluwer International Series in Engineering and Computer Science*, is a high level academic and professional series publishing research on the design and applications of analog integrated circuits and signal processing circuits and systems. Typically per year we publish between 5–15 research monographs, professional books, handbooks, edited volumes and textbooks with worldwide distribution to engineers, researchers, educators, and libraries.

The book series promotes and expedites the dissemination of new research results and tutorial views in the analog field. There is an exciting and large volume of research activity in the field worldwide. Researchers are striving to bridge the gap between classical analog work and recent advances in very large scale integration (VLSI) technologies with improved analog capabilities. Analog VLSI has been recognized as a major technology for future information processing. Analog work is showing signs of dramatic changes with emphasis on interdisciplinary research efforts combining device/circuit/technology issues. Consequently, new design concepts, strategies and design tools are being unveiled.

Topics of interest include:

Analog Interface Circuits and Systems;

Data converters;

Active-RC, switched-capacitor and continuous-time integrated filters;

Mixed analog/digital VLSI;

Simulation and modeling, mixed-mode simulation;

Analog nonlinear and computational circuits and signal processing;

Analog Artificial Neural Networks/Artificial Intelligence;

Current-mode Signal Processing; Computer-Aided Design (CAD) tools;

Analog Design in emerging technologies (Scalable CMOS, BiCMOS, GaAs, heterojunction and floating gate technologies, etc.);

Analog Design for Test;

Integrated sensors and actuators; Analog Design Automation/Knowledge-based Systems; Analog VLSI cell libraries; Analog product development; RF Front ends, Wireless communications and Microwave Circuits;

Analog behavioral modeling, Analog HDL.

More information about this series at http://www.springer.com/series/7381

Pietro Buccella • Camillo Stefanucci • Maher Kayal
Jean-Michel Sallese

Parasitic Substrate Coupling in High Voltage Integrated Circuits

Minority and Majority Carriers Propagation in Semiconductor Substrate

Springer

Pietro Buccella
STI IEL GR-KA
École Polytechnique Fédérale de Lausanne
Lausanne, Switzerland

Camillo Stefanucci
STI IEL GR-KA
École Polytechnique Fédérale de Lausanne
Lausanne, Switzerland

Maher Kayal
STI IEL GR-KA
École Polytechnique Fédérale de Lausanne
Lausanne, Switzerland

Jean-Michel Sallese
STI IEL EDALB
École Polytechnique Fédérale de Lausanne
Lausanne, Switzerland

ISSN 1872-082X ISSN 2197-1854 (electronic)
Analog Circuits and Signal Processing
ISBN 978-3-030-08976-4 ISBN 978-3-319-74382-0 (eBook)
https://doi.org/10.1007/978-3-319-74382-0

Printed on acid-free paper

This Springer imprint is published by the registered company Springer International Publishing AG part of Springer Nature.
The registered company address is: Gewerbestrasse 11, 6330 Cham, Switzerland

Foreword

In the early days, the high-voltage transistor was a discrete device, and the control and protection circuits were implemented with small ICs and discrete components. The interference between the circuits through common connections and supply lines was limited.

Due to rapid technology and packaging advancements, it became possible in the late seventies of the last century to integrate the high-voltage, high-power transistors with their protection and control circuits on the same chip: the Smart Power IC. This evolution is still ongoing with increasing power capability and switching speed of the power transistors and increasing accuracy and complexity of the protection and interface circuits on the same chip.

Most Smart Power ICs use junction isolation and increasingly suffer from interferences through the common substrate. Besides capacitive and inductive coupling between components, switching power devices also inject majority and minority carriers of considerable magnitude into the substrate. This gives rise to substrate biasing in the vicinity of the injection point and disturbance of sensitive interface circuits far away from the injection point.

The effects of these substrate currents are very difficult to predict since they propagate three-dimensionally through the substrate. Furthermore, traditional SPICE simulations only calculate with majority carriers and neglect minority carriers. Substrate couplings could hence not be included in the chip design cycle.

Traditionally, experimental test structures were used to quantify the substrate couplings in a given technology. Expert designers then devised circuit layout and protection structures with large margins to cope with the substrate interferences.

A much better and accurate approach is TCAD simulation of the whole chip. Such simulation includes the behavior of the minority carriers and the three-dimensional nature of the substrate current but requires tedious finite element modeling of the three-dimensional chip and lengthy simulations.

In this book, a new approach is presented. Equivalent electrical circuits are derived to model minority carriers in the substrate in SPICE, and a novel, meshing strategy is used to limit the number of nodes to model the chip layout. Together these advancements allow for simulation in SPICE of the parasitic substrate couplings in Smart Power ICs with adequate accuracy, and the simulation complexity is low enough to include critical layout interactions in the normal design cycle.

Herman Casier

Preface

Over the last century, electronics has been a major drive for the revolution in communication systems and in data computing. The demand for integrated circuits is accelerating as high integration of electronic functions is needed to share information and control our environment. The everlasting consumer demand for safety, energy efficiency, and connectivity explains why electronics is becoming of prime importance. In this context, the automotive industry is taking full advantage of this microelectronics revolution, and today, it happens to be one of the main users of integrated circuits.

Safety requirements for modern automotive electronics call for more and more robust circuits for power applications. In most situations, the mixed-signal circuit design flow cannot track ultimate design failures in complex power circuits mainly caused by parasitic substrate currents caused by majority/minority carriers. Traditionally, device simulations are carried out on a simplified model of the substrate to identify these parasitic signals. However, it is almost impossible to get accurate results with this approach. In addition, the circuit functionalities are excluded from these simulations since there is no back annotation between the substrate model and the circuit.

In this book, a novel substrate model and related extraction methodology is investigated. The substrate model consists in three generalized lumped elements—the EPFL diode, the EPFL resistor, and the EPFL homojunction—and a subdivision of the Integrated circuit (IC) layout into elementary cells where the continuity equation for the minority carriers is solved with the finite-difference method (FDM). A nonuniform mesh procedure is also implemented to minimize the number of nodes, giving rise to an equivalent three-dimensional model of the substrate. The concept is fully compatible with conventional circuit design tools, and it can be used for any high-voltage technology, including HV-CMOS and BCD processes. For instance, the single and parallel activation of substrate parasitic bipolar transistors in a $0.35\,\mu m$ High voltage (HV) technology is verified with SPICE simulations, predicting the activation of a latch-up in specific situations. To address the general problem of reliability in IC design, a fast method to monitor parasitic substrate currents is also presented. The substrate analysis and identification of critical substrate current path can be done during the preliminary design phase, allowing the placement of protections in an optimal layout floor plan.

This book can be used as reference for engineers and students designing HV ICs with high immunity to parasitic substrate current caused by the injection of minority and majority carriers. It is divided into seven chapters dealing with a specific aspect of the model or its applications.

Chapter 1 provides an overview of substrate coupling issues. The state of the art of modeling strategies, including minority and majority carrier propagation into the substrate, is presented, together with an overview of IC design flow showing the main open issues in parasitic coupling.

Chapter 2 focuses on the architecture of standard HV technologies, highlighting major causes and effects of parasitic substrate currents.

Chapter 3 details the mathematical derivation of the EPFL substrate model where equations are translated into equivalent enhanced devices in order to be solved by circuit simulators.

Chapter 4 analyzes the EPFL substrate model with respect to numerical device simulators, including DC, AC, transient, and temperature simulations. Finally, breakdown simulations of basic ESD devices are discussed and demonstrate the ability of the model to simulate snapback behaviors.

Chapter 5 describes the procedure developed to implement the substrate extraction tool still preserving back annotation with the circuit. A specific meshing and technology reduction methodology is also developed to simplify the extracted network.

Chapter 6 is the ultimate assessment of the model with respect to experimental data and discusses the performance in terms of simulation speed and memory storage.

Finally, in Chap. 7, a novel procedure is presented to simulate and control substrate currents aiming layout optimization. A survey of the most recent advances in designing passive and active isolation structures is also presented and analyzed in details with the model.

Lausanne, Switzerland

Pietro Buccella
Camillo Stefanucci
Maher Kayal
Jean-Michel Sallese

Acknowledgments

This work has been sponsored by the Swiss National Science Foundation (project no. 125321) and by the European commission under the European FP7 AUTOMICS project (FP7/ICT 314135). The authors would like to particularly thank Dr. Fabrizio Lo Conte for his initial contribution to this work as well as the partners of the European project AUTOMICS for their technical cooperation.

Contents

Acronyms

ADAS	Advanced driver assistance system
BCD	Bipolar CMOS and DMOS
BJT	Bipolar junction transistor
CAD	Computer-aided design
CAN	Controller area network
CB	Common base
CE	Common emitter
CMOS	Complementary metal-oxide semiconductor
DI	Dielectric isolation
DMOS	Double-diffused MOS
DPI	Direct power injection
DRC	Design rule check
DSP	Digital signal processor
DTI	Deep trench isolation
EC	European community
ECGR	Electron-collecting guard ring
ECU	Electronic control unit
EDA	Electronic design automation
EMC	Electromagnetic compatibility
EME	Electromagnetic emission
EMI	Electromagnetic immunity
EPFL	Swiss Federal Institute of Technology of Lausanne
ERC	Electrical rule checker
ESD	Electrostatic discharge
FA	Failure analysis
FBM	Finite box method
FDM	Finite-difference method
FEM	Finite element method
GLD	Generalized lumped devices

HBM	Human body model
HCGR	Hole-collecting guard ring
HF	High frequency
HS	High side
HV	High voltage
IC	Integrated circuit
ISO	International Organization for Standardization
JI	Junction isolated
KCL	Kirchhoff's current law
KVL	Kirchhoff's voltage law
LCM	Line centered mesh
LDMOS	Lateral diffused MOS devices
LOCOS	Local oxidation of silicon
LS	Low side
LuT	Look-up table
LVS	Layout versus schematic
MAAP	Multi-ring active analogic protection
MCC	Minority carrier circuit
MCM	Minority carrier mirror
MOS	Metal-oxide semiconductor
MOSFET	Metal-oxide semiconductor field effect transistor
μP	Microprocessor
NCM	Node centered mesh
NFA	Negative feedback activated
NMOS	Negative-channel metal-oxide semiconductor
NPN	Negative-positive-negative (transistor)
NUM	Nonuniform mesh
PDE	Partial differential equation
PDK	Process design kits
PIC	Power integrated circuit
PWM	Pulse-width modulation
RC	Resistance and capacitance
RF	Radio frequency
SCR	Silicon-controlled rectifier
SI	Self-isolated
SNS	Sah-Noyce-Shockley
SOA	Safe operating area
SOAC	Safe operating area check
SoC	System-on-chip
SOI	Silicon-on-insulator
SPT	Smart power technology
SRAM	Static random-access memory
SRH	Shockley-Read-Hall

SVE	Smart voltage extension
TCAD	Technology computer-aided design
TCC	Total current circuit
TLM	Transmission line model
TVS	Transient voltage suppressor
UM	Uniform mesh
VDMOS	Vertical diffused MOS

Chapter 1
Overview of Parasitic Substrate Coupling

1.1 Substrate Parasitic Current

In an IC, electrical couplings between analog and digital circuits take place through minimal impedance paths between the metal interconnections layers and the substrate. Concerning metals, the parasitic contributions generated by resistances and capacitances can be directly obtained from the layout. Concerning the semiconductor substrate, the situation is much more complex since both majority and minority carriers currents come into play and the minimum impedance paths are not easily identified.

Electrical parasitic coupling between analog and digital circuits in mixed-signal designs is commonly referred as *substrate noise*. In industrial and automotive applications, additional parasitic currents are generated by the switching of inductive loads or by stringent ESD and Electromagnetic compatibility (EMC) tests. For instance, a significant injection of minority carriers (electrons in a p-type substrate) in substrates occurs when power MOSFETs driving an inductive load are switched off. In this situation, the drain voltage of the power transistor is suddenly biased below the substrate potential, and the parasitic substrate Bipolar junction transistor (BJT) is activated.

The injected minority carriers generate a reverse substrate current that can be collected by n-type wells of digital and analog circuits nearby. Moreover the reverse current flow can be of hundreds of milliamperes and can generate a latch-up, which is the most disruptive effect that can be triggered by an injection of minority carriers.

In general, three kinds of substrate current can be identified, each of them having different characteristics and mitigation strategies [1, 2]. Here, a p-type substrate is considered.

- **Majority carriers current:** when a parasitic PNP transistor is activated, it injects majority carriers into the substrate, thus generating a substrate potential shift (de-biasing). The substrate acts as a distributed collector for the parasitic BJT,

P. Buccella et al., *Parasitic Substrate Coupling in High Voltage Integrated Circuits*, Analog Circuits and Signal Processing, https://doi.org/10.1007/978-3-319-74382-0_1

and the de-biasing becomes dependent on the equivalent bulk resistivity. Proper P+ grounding schemes are required to reduce these potential shifts;

- **Minority carriers current:** when the parasitic device is a NPN transistor, minority carriers are injected into the substrate and generate currents that interact with surrounding n-wells. Now, the substrate holds for the distributed base of a parasitic multi-collector BJT. The presence of additional n-wells and the layout floor planning have a huge impact in the resulting coupling currents;
- **Switching noise current:** AC currents from fast switching transistors can also couple through junction capacitances, generating high-frequency noise and substrate potential spikes. In this case the substrate acts as a distributed RC network, and the contact placements, as well as decoupling capacitors, are needed to mitigate the induced noise.

The first two types of substrate current are those generated in power stages of HV ICs when PN junctions of HV MOSFETs are forward biased and start injecting electrons (minority carriers) or holes (majority carriers) into the substrate. These charges can disturb nearby circuits and, in the worst case, can destroy the chip if a latch-up is triggered. Then, it is essential to adopt effective strategies at the circuit level (e.g., proper timing of transistor control signals) and during the layout stage (e.g., using appropriate guard rings and proper floor planning). However, no model is available to predict these parasitic signals during the design of the circuit.

Moreover, these kinds of couplings depend on the substrate doping as well. HV technologies require low doping in order to increase breakdown voltages. At the same time, low-doped substrates mean long minority and majority carrier's lifetimes and diffusion lengths. In this case, parasitic currents can generate coupling mechanisms far in the IC. To fix this issue and force recombination, a P++ highly doped substrate can be used together with a p-epi layer.

Concerning the switching noise generated during AC operation, it is attributed to capacitive coupling phenomena driven by the junction capacitances, in contrast to substrate current from bipolar transistors that are DC contributions. Switching noise is generated by high-frequency operation of digital and mixed-signal circuits, but still it is a source of noise for sensitive analog or RF cells [3]. This kind of substrate coupling can be easily modeled with a distribute RC network [4] as shown in Fig. 1.1 where a power stage, a logic block, and a sensitive bandgap circuit are placed nearby. Here the aggressor is the fast switching logic, and the noise detected depends on the equivalent impedance of the substrate [5]. But when the power stage is activated, meaning that a DC current sinks in a parasitic multi-collector structure, the RC modeling approach is unable to predict the induced shift in the bias voltage of the bandgap circuit (the selected victim) [6, 7]. In any case, during the switching of the power MOSFETs, AC and DC contributions coexist and must be taken into account.

In order to suppress substrate switching noise, the RC time constant must be increased by introducing grounding schemes and decoupling capacitances. Also in this case, the doping of the substrate has an influence since it fixes the resistance of the bulk [8]. For low-doped substrates, the resistivity is high, and the noise is

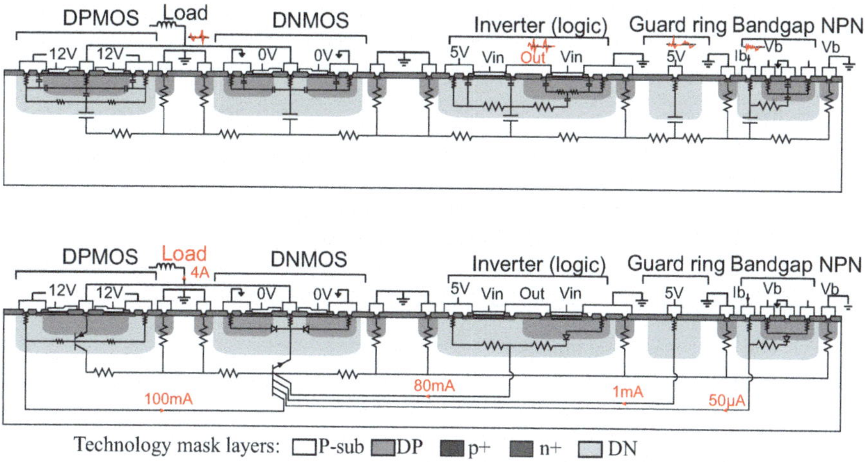

Fig. 1.1 Cross section showing impedance paths of substrate couplings: (top) switching noise from an inverter and (bottom) belowground condition of a H-Bridge injecting current into the substrate. In the first case, the n-well acts as a decoupling capacitor, while in the second case, it behaves as an additional collector for the parasitic NPN BJT device

decreased by increasing the distance between the noisy elements and the sensitive circuits. For high-doped substrates, the P++ bulk propagates the noise "uniformly" across the chip, and the distance has a negligible impact. However, in this case the resistivity is lower and also the overall RC noise.

1.2 Electrical Modeling of the Substrate: State of the Art

The coupling related to a parasitic NPN BJT is very dependent on the distance between emitting and collecting wells in the layout. There is no reliable compact model to simulate parasitic currents generated by these parasitic BJTs since the current gain β of a lateral NPN BJTs depends on the 3D layout configuration. Foundries may provide additional layout guidelines to avoid or reduce substrate coupling currents. Additionally, measurements of BJTs are done to estimate the sensitivity of parasitic signals to the layout [9, 10]. This is done by injecting a current I_E into the substrate and measuring the collected current I_C from different collector nodes. It results in some empirical fitting curves for the parameter $\alpha = I_C/I_E$ that is used to quantify the substrate couplings. Still, such results cannot be generalized for multi-collector or multi-emitter configurations due to the nonlinearity of minority carrier's current propagation. Finally, depending on the layout, additional coupling mechanisms are competing, and these are almost impossible to predict with a predefined BJT model.

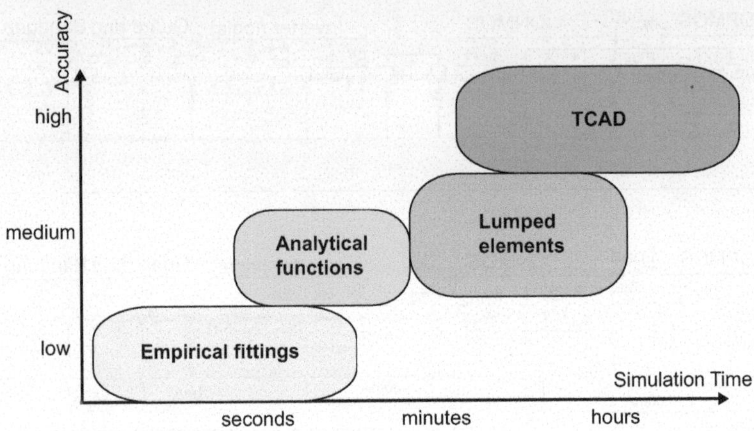

Fig. 1.2 Comparison of different substrate modeling methodologies

Different substrate modeling methodologies are compared in Fig. 1.2. The accuracy refers to the predictability of the model with respect to different layouts and electrical configurations. As expected, Technology computer-aided design (TCAD) software is the most accurate to simulate the substrate current since transport equations are solved numerically. However, this requires lots of computer resources and long simulation time and cannot be extended to a full IC. In this context, TCAD has been used to analyze substrate couplings in HV circuits like a H-Bridge [11] once the technology parameters are properly calibrated [12]. Then, co-simulation of mixed-signal design flow and TCAD as proposed by Gnani et al. [13] happens to be the most accurate solution to analyze substrate current coupling. Still, one drawback is the lack of back-annotation between the "selected" piece of substrate in TCAD and the original circuit schematic. A solution was proposed by Kollmitzer et al. [14] where TCAD is used to characterize minority carriers couplings and to create a model that can be included into the original circuit by the means of Look-up tables (LuTs). However, to ensure accuracy, a new TCAD simulation is required each time a different layout is analyzed.

Other numerical techniques have been exploited to reduce the simulation time down to seconds. One is to use Green's function to solve the diffusion equation of minority carriers [15]. This option requires a complex mathematical formulation that can hardly include drift currents, a serious limitation under high electric fields. Moreover, it cannot be integrated directly in circuit simulators.

This is the strategy adopted by Oehmen et al. [16]. In their approach the substrate is divided into spheres, and the diffusion equation is solved in polar coordinate satisfying Kirchhoff's laws. The equation is solved in each single sphere to calculate the injected electrons concentration, while the coupling currents are computed as linear combinations of all minority carriers' densities. Nevertheless, this methodology still focuses on the diffusion of minority carriers, neglecting drift current components and majority carriers (e.g., the vertical PNP BJT cannot be simulated with this approach).

Another concept was proposed by Lo Conte, Sallese, and Kayal [17] based on Cartesian coordinates compatible with the convention used for drawing IC layouts. The idea was to define two lumped components, namely, the enhanced diode and the enhanced resistor, that take into account the diffusion of the minority carriers at circuit level. In each lumped element, the analytical solution for the 1D diffusion equation is implemented with arbitrary boundary conditions. Next, these elements are interconnected, and the 3D network is computed while satisfying Kirchhoff's laws. It was shown that interconnection of these enhanced devices can effectively simulate substrate coupling currents in HV ICs [18]. In addition, they can be "linked" to the actual devices that are implemented in the circuit, a big step toward co-simulation of substrate *and* circuits in a single software. Nevertheless, drift phenomena are not taken into account, and its application is still limited to low-current/low-voltage regime.

High-current effects have been introduced by Stefanucci et al. [19] by extending Lo Conte's model to Generalized lumped devices (GLD). This approach solves the drift-diffusion equation in 3D, including contacts and AC operation. As of today this model, referred hereafter as the EPFL substrate model from the university where it was developed, is the most promising approach to simulate substrate current within circuit simulators. The model, which was validated in complex HV ICs, is detailed in the following chapters.

Finally, an historical note must be added since similar approaches were already proposed before TCAD software development, when semiconductor equations were solved through circuit simulators. Different equivalent circuits of drift-diffusion equations have been proposed in the literature, like Linvill's circuit [20] and Sah's circuit based on quasi-Fermi potentials [21]. Beside the intuitive derivation of these circuits, the mathematical basis is still close to one behind the EPFL model. However, Linvill's and Sah's circuits attempted to solve the complete system of drift-diffusion coupled equations. A similar, but more recent, diode model is the Laux-Hess equivalent circuit [22] where Poisson's equation and the minority/majority carriers continuity equations are solved simultaneously. Despite these models can be considered as initial TCAD software, today they are no longer competitive in regard to TCAD tools.

1.3 The Missing "Substrate" in IC Design Flow

If possible, failures in ICs should be predicted during the IC design phase. The typical design flow of any IC is shown in Fig. 1.3 and consists in different steps. Starting from electrical specifications given by the application, the system level architecture is determined first. At this stage each block of the system is designed at the transistor level using the selected technology and related SPICE models. Several Electronic design automation (EDA) tools are available to simulate circuits, including temperature, process mismatch, and noise degradation effects. Once the circuit behaves as expected, the layout is finalized following the Process design

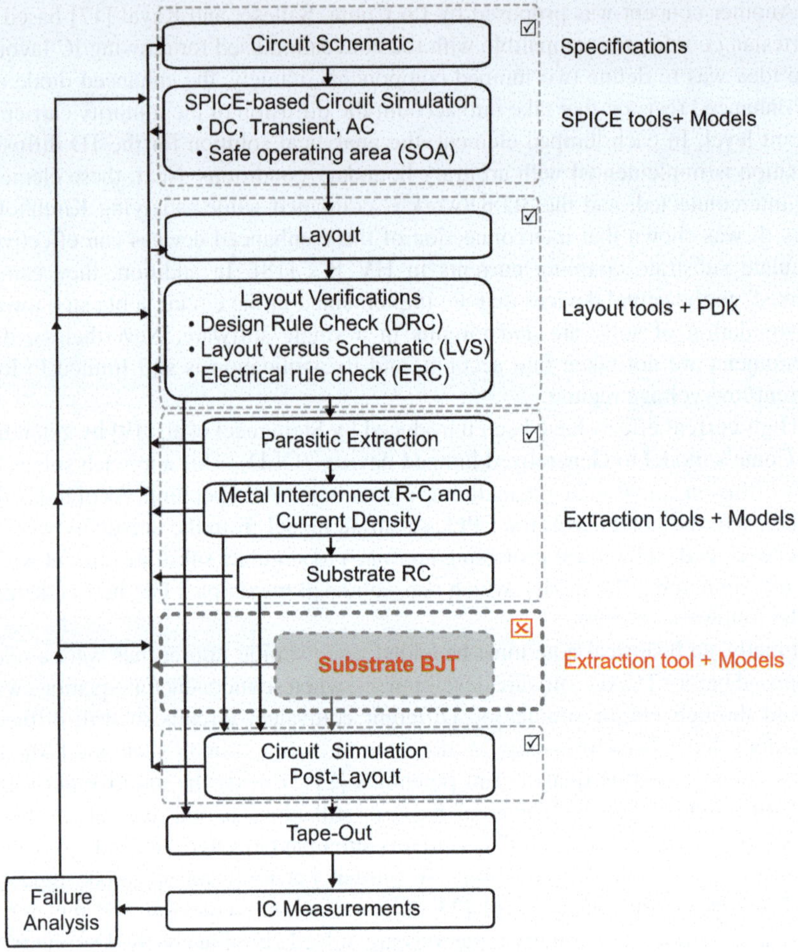

Fig. 1.3 Simplified IC design flow. The missing substrate parasitic extraction step with BJT transistors is highlighted

kits (PDK) rules provided by the foundry. When the Design rule check (DRC) and Layout versus schematic (LVS) tests are passed, parasitic capacitances of metal interconnections introduced by the layout are estimated. Many commercial tools allow the extraction of RC parasitic components starting from the IC mask layers, enabling reliable post-layout simulations. If the circuit performance is not degraded by the layout, it can be sent to tape-out. On the other hand, if a problem arises, one or more steps back are required to fix the issue. Any reiteration during the design/redesign process has a direct impact on design costs.

If testing the product reveals that specifications are not met, or if an electrical failure is evidenced, an investigation process is required to detect the origin of the dysfunction. Failures can be categorized in three groups: hard failure (destruction), latent failure (not working after certain time), and soft failure (degradation of electrical properties). The Failure analysis (FA) is widely used to understand the nature of the problem and to localize the region involved in the fault. Unfortunately, this FA is very expensive and frequent in case of HV ICs or ESD testing.

It is clear that if simulations can predict such failures, costly redesign can be avoided. A solution could be to run a final verification using specialized physics-based software. This option often relies on TCAD tools that enable co-simulation with mixed-mode circuit analysis. Today this approach is used for electrothermal phenomena [23], minority carriers related propagation effects [13], and ESD protection design [24, 25]. An accurate description of the technology is then required to set up TCAD simulations, but this information is often not available for design engineers. Moreover designers prefer to have a circuit based approach, which is more intuitive and more effective in terms of simulation time. But since no such a tool exists, usually designers still rely on empirical rules and test structures with different layout, a method that suffers from costly trial-and-error implementations.

Despite research efforts toward TCAD based solutions, from the point of view of reliability issues analysis, a step in the design flow is clearly missing: the simulation of the substrate with parasitic BJTs. Such a block can contribute to identify failures during HV IC design as it gives a complete description of electrical couplings inside the IC and can help designers to simulate cross couplings and optimize the circuit layout. However, the substrate model cannot be implemented at the schematic level because coupling mechanisms are strongly layout dependent. In the same spirit, device compact models neglect the layout and the minority carriers that are fundamental to simulate the influence of the substrate. To fix these reliabilities issues, a layout-aware substrate model must be included along the design flow.

Nowadays the available substrate models are only limited to mixed-signal switching noise. Many tools were developed to extract RC parasitic components of the substrate, depending on the technology doping profiles [26, 27]. Optimization techniques are also exploited to reduce the parasitic netlist and to compute the equivalent impedance in the circuit layout [28, 29]. The same model based on RC components is used to propagate currents and voltages across the chip and to estimate ESD effectiveness [30]. Nevertheless, there is still no substrate tool available that takes into account parasitic BJT and minority carriers.

The development of such a SPICE-compatible substrate modeling tool is the aim of this book.

References

1. A. Hastings, *The Art of Analog Layout* (Prentice Hall, Upper Saddle River, 2005)
2. R.J. Widlar, Controlling substrate currents in junction-isolated ICs. IEEE J. Solid State Circuits **26**(8), 1090–1097 (1991)
3. A. Abdel-Ghaffar, M. Ismail, *Substrate Noise Coupling in RFICs* (Springer Science & Business Media, New York, 2008)

4. T.A. Johnson, R.W. Knepper, V. Marcello, W. Wang, Chip substrate resistance modeling technique for integrated circuit design. IEEE Trans. Comput. Aided Des. Integr. Circuits Syst. **3**(2), 126–134 (1984)
5. B. Nauta, G. Hoogzaad, Substrate bounce in mixed-mode CMOS IC's, in *Workshop on Substrate Noise Coupling in Mixed-Signal IC's* (1998), pp. 201–213
6. I. Zheng, H. Zhao, J. Yang, W. Li, W. Li, L. Wei, C. Wang, Analysis of a bandgap circuit DC shift caused by substrate noise generated by power switches, in *8th International Conference on Solid-State and Integrated Circuit Technology (ICSICT)*, Oct 2006, pp. 1718–1720
7. W. Horn, H. Zitta, A robust smart power bandgap reference circuit for use in an automotive environment, in *Proceedings of the 27th European Solid-State Circuits Conference (ESSCIRC)*, Sept 2001, pp. 217–220
8. D.K. Su, M.J. Loinaz, S. Masui, B.A. Wooley, Experimental results and modeling techniques for substrate noise in mixed-signal integrated circuits. IEEE J. Solid State Circuits **28**(4), 420–430 (1993)
9. B. Murari, F. Bertotti, G.A. Vignola, A. Andreini, *Smart Power ICs: Technologies and Applications* (Springer-Verlag GmbH, Berlin, 1996)
10. O. Gonnard, G. Charitat, P. Lance, E. Stefanov, M. Suquet, M. Bafleur, N. Mauran, A. Peyre-Lavigne, Substrate current protection in smart power IC's, in *Proceedings of the IEEE 12th International Symposium on Power Semiconductor Devices and ICs (ISPSD)* (2000), pp. 169–172
11. M. Schenkel, Substrate current effects in smart power ICs. PhD thesis, ETH Zürich, Nr. 14925, 2003
12. M. Schenkel, P. Pfaeffli, W. Wilkening, D. Aemmer, W. Fichtner, Transient minority carrier collection from the substrate in smart power design, in *Proceedings of the 31st European Solid-State Device Research Conference (ESSDERC)*, Sept 2001, pp. 411–414
13. E. Gnani, V. Giudicissi, R. Vissarion, C. Contiero, M. Rudan, Automatic 2-D and 3-D simulation of parasitic structures in smart-power integrated circuits. IEEE Trans. Comput. Aided Des. Integr. Circuits Syst. **21**(7), 791–798 (2002)
14. M. Kollmitzer, M. Olbrich, E. Barke, Analysis and modeling of minority carrier injection in deep-trench based BCD technologies, in *9th Conference on Ph.D. Research in Microelectronics and Electronics (PRIME)*, June 2013, pp. 245–248
15. M. Corradin, A. Sangiovanni-Vincentelli, E. Charbon, Modeling minority carrier diffusion through substrate in SMART power ICs, in *IEEE Behavioral Modeling and Simulation (BMAS)* (2005)
16. J. Oehmen, L. Hedrich, M. Olbrich, E. Barke, A methodology for modeling lateral parasitic transistors in smart power ICs, in *IEEE Behavioral Modeling and Simulation (BMAS)*, Sept 2005, pp. 19–24
17. F. Lo Conte, J.-M. Sallese, M. Kayal, Smart power IC simulation of substrate coupled current due to majority and minority carriers transports, in *IEEE International Conference on IC Design and Technology (ICICDT)*, June 2010, pp. 168–171
18. F. Lo Conte, J.-M. Sallese, M. Kayal, Modeling methodology of high-voltage substrate minority and majority carrier injections, in *Proceedings of the European Solid-State Device Research Conference (ESSDERC)*, Sept 2010, pp. 194–197
19. C. Stefanucci, P. Buccella, M. Kayal, J.-M. Sallese, Spice-compatible modeling of high injection and propagation of minority carriers in the substrate of smart power {ICs}. Solid State Electron. **105**, 21–29 (2015)
20. J.G. Linvill, Lumped models of transistors and diodes. Proc. IRE **46**(6), 1141–1152 (1958)
21. C.T. Sah, The equivalent circuit model in solid-state electronics-III: conduction and displacement currents. Solid State Electron. **13**(12), 1547–1575 (1970)
22. S.E. Laux, K. Hess, Revisiting the analytic theory of p-n junction impedance: improvements guided by computer simulation leading to a new equivalent circuit. IEEE Trans. Electron Devices **46**(2), 396–412 (1999)

23. A.-G. Bajenaru, C. Boianceanu, G. Brezeanu, Investigation of electro-thermal behaviour of a linear voltage regulator and its protection circuits by simulator coupling, in *International Semiconductor Conference (CAS)*, vol. 2, Oct 2013, pp. 237–240
24. A.Z. Wang, C.H. Tsay, A new design methodology using simulation for on-chip ESD protection designs for integrated circuits, in *5th International Conference on Solid-State and Integrated Circuit Technology* (1998), pp. 509–512
25. H. Feng, G. Chen, R. Zhan, Q. Wu, X. Guan, H. Xie, A.Z.H. Wang, R. Gafiteanu, A mixed-mode ESD protection circuit simulation-design methodology. IEEE J. Solid State Circuits **38**(6), 995–1006 (2003)
26. F.J.R. Clement, E. Zysman, M. Kayal, M. Declercq, LAYIN: toward a global solution for parasitic coupling modeling and visualization, in *Proceedings of the IEEE Custom Integrated Circuits Conference (CICC)*, May 1994, pp. 537–540
27. I.L. Wemple, A.T. Yang, Integrated circuit substrate coupling models based on Voronoi tessellation. IEEE Trans. Comput. Aided Des. Integr. Circuits Syst. **14**(12), 1459–1469 (1995)
28. K.J. Kerns, I.L. Wemple, A.T. Yang, Stable and efficient reduction of substrate model networks using congruence transforms, in *IEEE/ACM International Conference on Computer-Aided Design (ICCAD)*, Nov 1995, pp. 207–214
29. A. Koukab, K. Banerjee, M. Declercq, Modeling techniques and verification methodologies for substrate coupling effects in mixed-signal system-on-chip designs. IEEE Trans. Comput. Aided Des. Integr. Circuits Syst. **23**(6), 823–836 (2004)
30. Mentor Graphics. Calibre® perc™

Chapter 2
Design Challenges in High-Voltage ICs

2.1 Smart Power ICs

The growth and development of power electronics are closely related to the progress in developing power semiconductor devices. Initially, power systems were limited to discrete control circuits and power devices. However, the need for compact and flexible solutions with improved performances led to the development of specific HV technologies to build monolithic circuits with digital, analog, and power devices. As a result, new families of HV ICs and technologies were introduced to improve the quality and implementation of power electronics. Various terminologies are used to define these HV ICs and technologies: Power integrated circuit (PIC), HV ICs, Power ICs, Smart Discrete ICs, and Smart Power ICs. The terminology *Smart Power IC* is widely accepted and used to refer to such HV circuits and technologies. However, in this book, these terms are equally used.

Main markets for Smart Power ICs concern the automotive and industrial applications that evolved drastically over the last few decades. From limited circuits using a single power switch with control and protection circuitry [1], modern designs include complex System-on-chip (SoC) with a Digital signal processor (DSP) along with the power devices [2]. Figure 2.1 shows a typical block diagram of a modern Smart Power IC, including power devices in addition to digital and analog circuits [3, 4].

Among all available power devices, the Double-diffused MOS (DMOS) transistor is currently the most used in Smart Power ICs [5]. Moreover, extra analog and digital circuits add self-protection and diagnostic functions, thus making the DMOS particularly reliable and effective [6]. Indeed, these digital and analog circuits build the smart components of ICs.

Self-protection features can sense situations which are beyond the typical operating conditions of the circuit, such as overvoltage, overcurrent, and overtemperature conditions. On the other hand, circuits for diagnostic features such as no-load and undervoltage are implemented for power management purposes and

P. Buccella et al., *Parasitic Substrate Coupling in High Voltage Integrated Circuits*, Analog Circuits and Signal Processing, https://doi.org/10.1007/978-3-319-74382-0_2

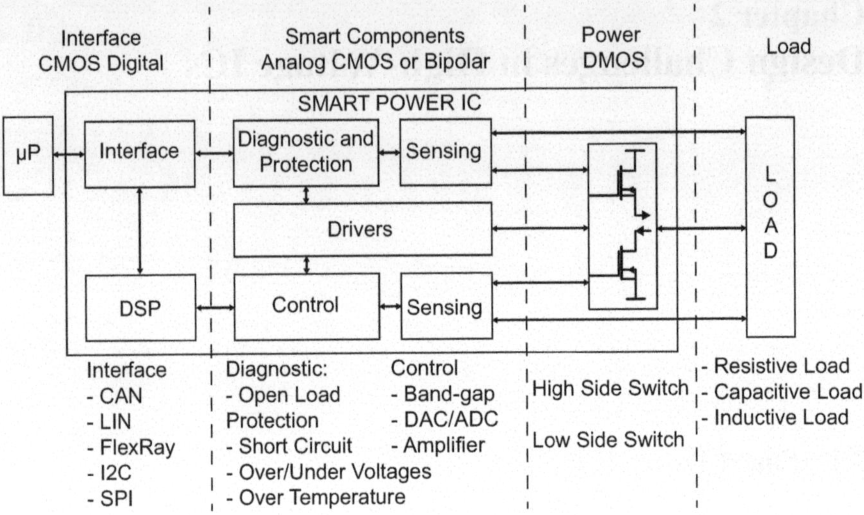

Fig. 2.1 Typical block diagram of a Smart Power IC reproduced from [3]

to improve the system reliability. These fault and diagnostic conditions are fed via status signals to logic circuits transfer the information to the outside world with well-known communication protocols (i.e., SPI, UART, I2C, CAN, LIN, FlexRay).

In many Smart Power ICs, additional precision analog circuits are required for the implementation of the control loop circuitries, such as bandgap voltage references, oscillators, ramp generators, current generators, and amplifiers. For instance, such precision blocks are used in power management circuits such as battery chargers and in power supply systems for on board or on-chip power conversion.

In Smart Power IC, a significant gap exists between the voltage in power and digital domains. The typical high-voltage range for power applications is between 12 and 120 V, while the low-voltage circuitry is in the 1–5 V range. Moreover, HV circuits are getting larger to deliver more power, while low-voltage circuits are shrinking and become more sensitive to parasitic couplings. The simple rule of separating high-voltage and low-voltage domains does not always ensure a safe operation of the IC. Therefore, special care must be taken with the IC layout to isolate the control circuitry from power devices in order to minimize parasitic couplings. This aspect adds many technical challenges to the IC design and pushed semiconductor companies to develop specific HV technologies optimized in terms of cost, electrical performances, and especially electrical isolation against substrate couplings.

2.2 High-Voltage Technologies

HV technologies are divided in bulk and SOI technologies [7]. Bulk technologies use silicon as the bulk material. Further classification identifies power devices where terminals shared between the top and bottom sides of the wafer and power devices where terminals are implemented on top. For the first case, these technologies are known as vertical processes, while for the second case, they are known as lateral processes. SOI technologies on the other hand use SOI wafer as the starting material.

2.2.1 *Vertical Bulk Technologies*

In vertical technologies, the Vertical diffused MOS (VDMOS) or vertical bipolar transistors represent the basic power device. They are also known as common-drain technologies, where the wafer is the drain of the power devices. Historically, these processes were derived from discrete power technologies where Complementary metal-oxide semiconductor (CMOS) and HV lateral devices were added for additional control.

The flow of high currents is perpendicular to the wafer surface as the drain of the power device is on the backside of the wafer. The cross section of a common-drain process is shown in Fig. 2.2a. Here, the n-type substrate is the drain. The process is simple and cheap since few additional masks are needed to include control devices. However, since the substrate is common to all devices, circuit configurations are limited to common-drain configurations and restrict this technology to simple power circuits. For example, discrete HV switches designed to handle large currents of up to 10 A or more are built on such vertical technologies [8, 9].

2.2.2 *Lateral Bulk Technologies*

Lateral bulk technologies include HV-CMOS and BCD processes. As a common characteristic, these processes have a similar structure and combine digital, analog, and power devices on the same surface of the wafer [10]. This gives more flexibility when interconnecting multiple power transistors in High side (HS) and Low side (LS) drivers on the same IC.

HV-CMOS processes are based on low-voltage CMOS technology with the addition of extra masks to extend the voltage capability of MOS transistors [11]. The simplified cross section of a lateral HV-CMOS process is shown in Fig. 2.2b.

Examples of BCD technologies are shown in Fig. 2.2c, d. BCD technologies combining analog, digital, and power devices in one process platform were introduced for the first time in the mid-1980s by SGS Thomson [12]. The starting

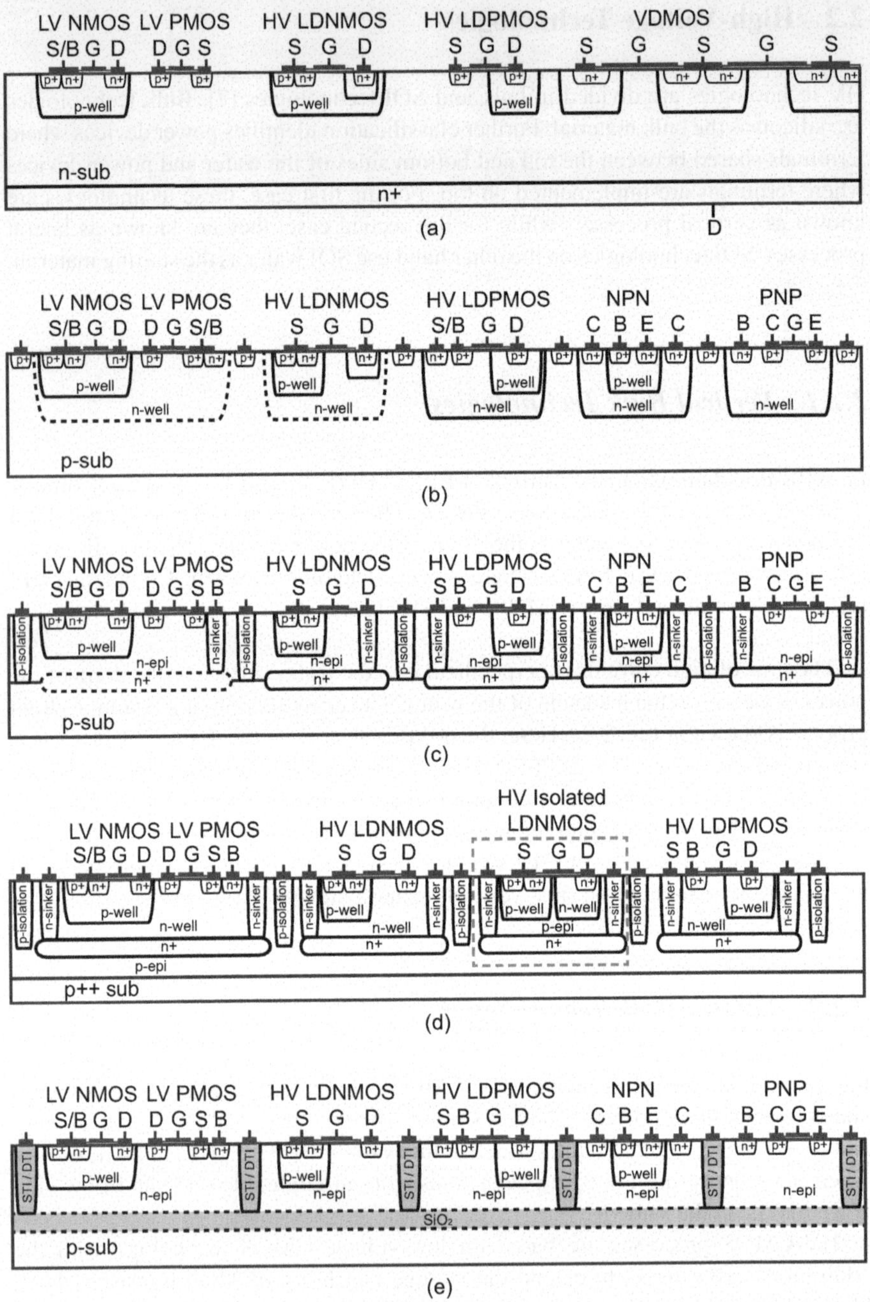

Fig. 2.2 Cross section of main smart power technologies: vertical bulk based (**a**), lateral bulk HV-CMOS (**b**), n-epi p-substrate (**c**), and p-epi p-substrate BCD based. SOI and DI based (**e**)

material is a p-type wafer with a n-type or p-type epitaxial layer. BCD technologies use highly doped n-type buried layers and n-sinker. Those layers allow for the implementation of NPN and PNP bipolar transistors with high analog precision. In addition, they increase isolation between the devices. Figure 2.2c illustrates an example of a BCD process where a n-type epitaxial layer is grown on a p-substrate to accommodate various devices, isolated by vertical and deep p-type regions. In Fig. 2.2d is another example of a BCD process with a p-type epitaxial layer. Often the wafer is a highly doped p++ substrate. Furthermore, this process architecture offers as an option an isolated NDMOS [13–15].

2.2.3 *SOI and DTI Technologies*

SOI processes use the SOI as the starting material. As shown in Fig. 2.2e, the silicon oxide is used to isolate vertically and horizontally each device. Similar to BCD, they combine bipolar, CMOS, and DMOS implemented in active silicon layers on top of the buried oxide.

DTI is an alternative to the SOI isolation. The process flow is similar to the conventional BCD. Silicon is used as the starting material, and undoped polysilicon is used to fill deep trenches to isolate the devices laterally. The technology cross section of the DTI process is similar to Fig. 2.2e, except that the buried oxide layer is not present.

2.3 Substrate Parasitic Transistors

The typical substrate parasitic devices consist of lateral NPN and vertical PNP bipolar transistors. With exception to SOI technologies, such parasitic structures are present in each bulk technologies as shown in Fig. 2.2 from (a) to (d), including the DTI process. However in DTI processes, the deep isolation reduces substrate couplings by several orders of magnitude [16].

Substrate parasitic BJT devices in HV-CMOS, BCD with n-epi, and BCD with p-epi on a highly doped substrate (p++) are shown in Fig. 2.3. NPN transistors are lateral devices having the common substrate as base and n-wells as either emitters or collectors. The PNP transistor is on the other hand a vertical device with the common substrate acting as the collector. The emitter can be any p-well enclosed in an n-type well, and the collector is the substrate. Both parasitic structures are activated when these two specific conditions are met [7]:

- The lateral NPN transistor is turned on when the potential of one or multiple n-wells is pulled down below the substrate potential or when a current is drawn from a n-well. The n-well(s) acts as an emitter, and any other n-well biased at a voltage greater or equal to the substrate potential acts as a collector of the NPN device.

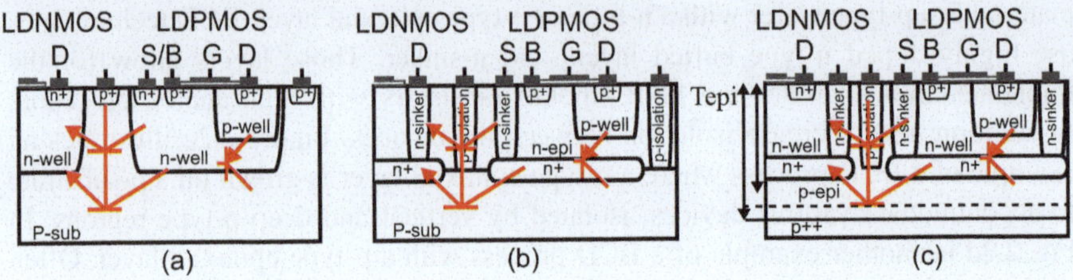

Fig. 2.3 Substrate parasitic BJT devices: lateral NPN and vertical PNP transistors in HV-CMOS (**a**), BCD with n-epi (**b**), and BCD with p-epi/p++ (**c**) processes

- The vertical PNP transistor is turned on when a p-type well is biased above the voltage of its enclosing n-well or when a current is injected in a p-well enclosed in a n-well. In this case, the whole substrate is the collector.

However, a NPN transistor can lead to the activation of a PNP transistor, and similarly a PNP can activate a NPN. This cumulative effect can trigger a latch-up.

2.3.1 Choice of the Technology

Currently, the different HV technologies are in competition for their applications in automotive and industrial electronics [17]. They all provide power devices with similar electrical characteristics such as low on-resistance and high current ability. Most recent HV-CMOS processes compete with state-of-the-art BCD processes in terms of low on-resistance [18]. Often, the choice for one or another process is cost-driven.

However, in HV applications, substrate parasitic bipolar structures are regularly activated under specific conditions, and their effects must be considered from the early stages of design. The electrical characteristics of such parasitic transistors are strongly technology-dependent, and the choice for a technology should rely not only on electrical performances and economical aspects but also on the isolation features that they provide. Designers must be aware of the electrical characteristic of the parasitic bipolar structures early in the design process. Table 2.1 shows a comparison of wafer costs with different isolation methods [19].

The isolation between devices in bulk technologies (Fig. 2.2a–d) relies on reverse-biased PN junctions. For this reason they are also classified as Junction isolated (JI) processes. Among HV-CMOS and BCD processes, the starting material is either a low-doped or a high-doped p-type substrate.

In a HV-CMOS technology (Fig. 2.2b), the low-doped substrate enhances the gain of the lateral NPN transistor. This is mainly due to the high diffusion length of minority carriers (electrons) into the substrate (NPN base). The gain of the vertical PNP is also relatively high partly because no heavily doped n-layers are available.

Table 2.1 Comparison of HV technologies with respect to wafer costs and isolation methods

Isolation strategy	Substrate	Lateral NPN gain	Vertical PNP gain	Wafer cost
Junction isolation	HV-CMOS low-doped substrate	High	High	Low
	BCD with n-epi over low-doped substrate	Moderate with low injected currents	Low	Potentially low
	BCD with p-epi over high-doped substrate	Moderate with low injected currents	Low	Potentially low
DTI	BCD with p-epi over high-doped substrate	Very low if DT reaches p++	Low	High
SOI	SOI	NPN nonexistent	PNP nonexistent	High

In BCD technologies, thanks to the additional high-doped n-buried and n-sinker layers, both parasitic structures have a lower gain. In BCD architectures, based either on n-epi or p-epi (Fig. 2.2c, d), the inclusion of the n+ buried layer degrades the current gain (β) of the parasitic PNP transistor below unity, even at high temperatures. The diffusion length of minority carriers (holes) into the highly doped n-layer (base of the PNP) is minimized, resulting in a reduced gain of the vertical PNP transistor [20]. Similarly, the gain of the lateral NPN transistor between n-epi islands is lowered by the highly doped n-type buried layer. The n-sinkers provide a low-ohmic preferential path for currents, thus avoiding their spreading into the substrate [20]. In some BCD processes, the p-epitaxial layer has a doping profile with a gradient toward the highly doped p++ substrate. This technique lowers even more the gain of the lateral substrate NPN transistor but is also dependent on the thickness of the epitaxial layer [21].

Technology solutions like SOI and DTI [22] are the preferred choice in most applications where reducing the substrate currents is mandatory. Nevertheless, these solutions are much more expensive than standard JI technologies. However, the higher parasitic capacitances of the SOI substrate may equally trigger a latch-up by means of transient couplings [23]. JI technologies remain the most cost-effective and preferred for the design of HV ICs in most applications. Compared to SOI and DTI, a deep understanding of the electrical characteristics of substrate BJT transistors is necessary to achieve a high degree of isolation between the multiple circuit devices.

2.4 Causes of Substrate Currents in High-Voltage ICs

High-voltage ICs are often used in applications that require to drive inductive loads, such as solenoids, relays, and motors. This category also includes power switching regulators, where the inductor is used as an energy storage element. Whenever an inductive load is switched off, substrate parasitic devices may be activated, and the current stored in the inductor may flow into the substrate.

Besides the control of inductive loads, ICs for automotive electronics must meet the safety standards released by the International Organization for Standardization (ISO) and automotive manufacturers. Integrated circuits for the automotive industry undergo additional stress tests where substrate current is generated by external electrical events such as high-voltage pulses. In the following sections, two situations in which substrate parasitic devices are activated are discussed:

- switching inductive load;
- automotive EMC requirements.

2.4.1 *Switching Inductive Load*

Electrical motors and solenoids are classical examples of electromechanical devices that are activated by a current flowing through their inductive components. The equivalent circuit of an electromechanical device consists mainly of an inductor (with inductance L) and a small series resistor with resistance R_L, with $R_L \ll L$. Such a circuit behaves mainly as an inductive load.

In an inductive load, the current is controlled by two driving phases which are called *storage* and *decay* phases. During the storage phase, the inductor is powered by a positive constant voltage source (i.e., battery voltage V_{BAT}) through a switch. With reference to the circuit in Fig. 2.4a, when the switch is on, the inductor current gradually builds up, thus storing energy in the magnetic field. According to Lenz's law, the magnetic field induces a voltage drop across the inductor v_L, and the differential equation during the inductor L_L charge phase is

$$L_L \frac{di_l}{dt} + (R_L + R_S)i_L = V_{BAT} \tag{2.1}$$

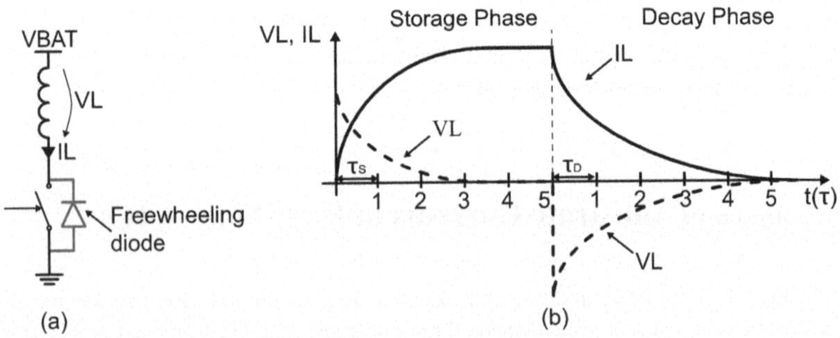

Fig. 2.4 Load voltage and current during storage and decay phases

where V_{BAT} is the battery voltage, R_L is the inductive load parasitic resistance, and R_S is the switch on-resistance. Solving (2.1) with $i_L(0) = 0$ as the initial condition, the time-dependent inductor current is obtained:

$$i_L(t) = \frac{V_{BAT}}{R_S + R_L}\left(1 - e^{-\frac{t}{\tau_S}}\right), \quad \text{with } \tau_s = \frac{L_L}{R_S + R_L} \tag{2.2}$$

τ_s is the rise time constant of the inductor current. At $t \approx 5\tau_s$ (Fig. 2.4), the current reaches the maximum steady value as determined by the inductor series resistance:

$$I_m \sim \frac{V_{BAT}}{R_S + R_L}. \tag{2.3}$$

At this point the storage phase is over and the inductor is equivalent to a short circuit. The energy stored in the inductor magnetic field is

$$U = \frac{1}{2}L_L I_m^2 \tag{2.4}$$

During the decay phase, the switch is turned off, and the inductor current must decrease instantaneously from Im to zero at high $\frac{di_L}{dt}$. In reality the inductor voltage v_L reverses its polarity and drops instantaneously from zero to a negative value, thus generating a voltage spike whose amplitude depends on the current slope and the inductance value:

$$v_L(t) = L\frac{di_L}{dt} \tag{2.5}$$

The transient curves for the inductor storage and decay phases are shown in Fig. 2.4b. The switch must be able to absorb rapidly the energy stored in the inductor and handle the voltage spike. Typical driving currents are in the order of a few amperes. If the desired inductor current discharge rate is 0.1 A/μs, the drain voltage rises to 100 V for a 1 mH inductor, which in some cases is critical as those values can surpass the technology breakdown voltage.

To protect the switch, the voltage spike must be carefully controlled. The decay time and inductor voltage amplitude are safely controlled by providing an alternate path to the decay current. In discrete applications, the alternate path is provided by external circuitry, such as freewheeling diodes, to protect the switch. Through the alternate path, which is commonly referred to as a freewheeling path, the inductor current decreases gradually from its previous steady value to zero, and the energy stored in the inductor magnetic field is released.

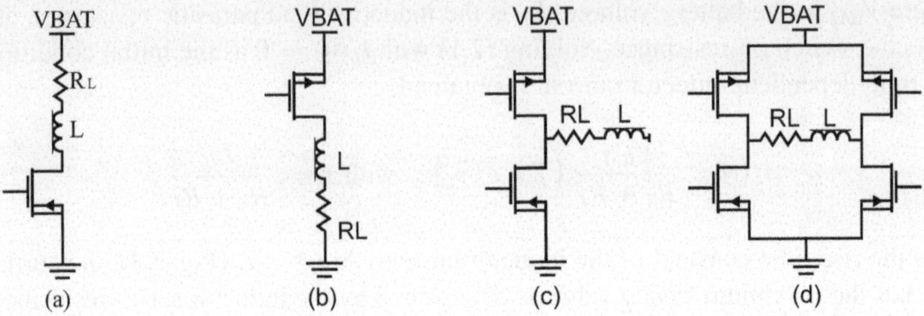

Fig. 2.5 Power switch configurations for driving and inductive load. Low-side (**a**), high-side (**b**), Half-Bridge (**c**), and H-Bridge (**d**) drivers

2.4.1.1 Controlling Decay Phase in High-Voltage Drivers

Common circuit architectures for driving inductive loads use power switches to control the storage and decay phase. The most common high-voltage drivers are shown in Fig. 2.5 and include LS, HS, Half-Bridge, and Full H-Bridge configurations. Each architecture has its own advantages and disadvantages, but most importantly, each architecture must handle overvoltages and undervoltages generated by the inductor current. Such voltage spikes can surpass the capabilities of the switch devices, and they can induce substantial injection of current into the IC substrate.

To assure circuit reliability, the voltage transients due to switching inductive loads must be kept within the maximum voltage and current ratings of devices. In integrated solutions, the inherent body diode of a DMOS is used to provide a freewheeling path to the inductor current during the decay phase. The use of the body diode of a DMOS as a freewheeling path, on one hand, eliminates external components and reduces the cost. On the other end, it generates a flow of parasitic substrate current in circuits built on JI technologies. Examples of Half- and Full H-Bridge are presented to illustrate substrate current generation and propagation.

Half-Bridge Configurations The configuration shown in Fig. 2.5c is often referred to the Half-Bridge circuit and is one of the most common circuit topologies used in power electronics. It is used in various applications such as synchronous power converters and motion control. It consists of two DMOS transistors arranged in a push-pull configuration.

The output of the Half-Bridge driver is continuously switched between the power supply, referred to as V_{BAT}, and the ground to control the current in the inductive load. Note that the switch driver provides cross conduction protection in order to avoid a shoot-through between V_{BAT} and the ground through the power switches. The cross conduction protection is implemented by introducing a dead time in the driver circuit, preventing the top and bottom switches from turning on simultaneously. During the dead time, the inductor current flows through the body diode of either the HS or the LS switch, depending on whether or not the other terminal of the inductor is high or low. When the second inductor terminal is

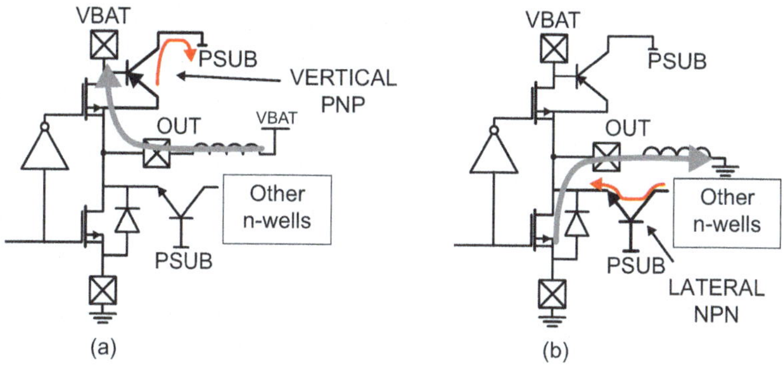

Fig. 2.6 Activation of the parasitic vertical PNP (**a**) and lateral NPN (**b**) in a Half-Bridge circuit made on junction-isolated HV technologies

connected to V_{BAT}, the current flows into the HS switch body diode during the dead time. In JI HV technologies, this current also turns on the parasitic vertical PNP transistor as shown in Fig. 2.6a. On the other hand, when the other inductor terminal is connected to ground, the currents flow in the LS body diode, also turning on the parasitic lateral NPN transistor (Fig. 2.6b).

In any case, a parasitic current flows into the substrate. The injected current is highly dynamic and depends on the dead time duration. In power management integrated circuits, dead time is optimized to minimize losses, especially at high loads. However at lighter loads, when the synchronous rectification technique is not used to increase the system efficiency (i.e., one of the two switches is off during the decay phase), a larger part of the inductor current is injected into the substrate. The impact of such a current may cause unforeseen circuit malfunction, and an accurate analysis of the effects of parasitic BJTs should be estimated during the design phase.

H-Bridge Configurations The H-Bridge consists of four DMOS transistors used as power switches to drive an inductive load. In the example shown in Fig. 2.7, the H-Bridge includes two n-type LS switches (N1 and N2) and two lateral high switches (P1 and P2). The switches are controlled by non-overlapping gate drivers which selectively turn on certain DMOS for four quadrant motor control applications [24].

A dead time is introduced in the synchronous driving circuitry of the switches in order to prevent a large current punch-through between the top and bottom switches. During the dead time phase transitions, the inductor current turns on the intrinsic body diode of either the bottom or top switch, depending on the direction of the inductor current. During this phase, also called the freewheeling phase, the H-Bridge output voltages (voltages at OUT1 or OUT2 in Fig. 2.7) go "belowground" or "above supply" voltage levels, thus injecting currents into the substrate when the circuit is built on a HV technology using junction isolation. The injected substrate current activates a parasitic bipolar transistor that can increase

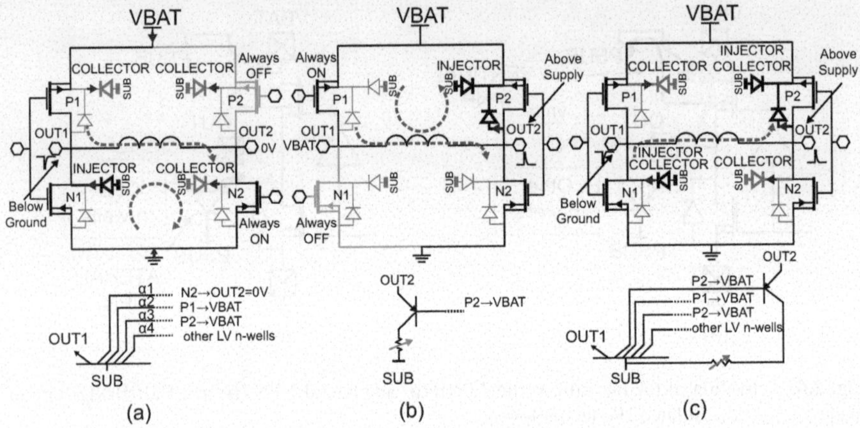

Fig. 2.7 Basic topology of an H-Bridge output power stage. Diagram of the forward-freewheeling transition, where OUT2 undergoes GND voltage (**a**). Diagram of the reverse-freewheeling transition, where OUT2 exceeds VBAT voltage (**b**). Diagram of the forward-reverse transition, OUT1 undergoes GND voltage, and OUT2 exceeds VBAT voltage (**c**)

circuit power dissipation, degrade circuit performances, or even destroy the chip if a latch-up is triggered. To avoid failure risks, typical H-Bridge applications recommend using external Schottky diodes as shunting devices in parallel with the power switches. Alternatively, when an H-Bridge is operated in asynchronous mode, the switch body diodes conduct the full load current, while it decays.

Depending on the control scheme of the switches and the inductor current direction, an H-Bridge can control three main operation modes: forward-freewheeling, reverse-freewheeling, and forward-reverse. In each of them substrate BJT transistors are activated as follows [25]:

(a) FORWARD-FREEWHEELING. This operating mode is shown in Fig. 2.7a. The HS switch P2 is always OFF and N2 is always ON. P1 and N1 are switched ON one at a time to increase or decrease the inductor current (i.e., accelerate or brake the motor). During the P1/N1 dead time transition, the inductor current forces the N1 drain voltage to go belowground.

    The parasitic N1 PN diode becomes forward biased and injects a parasitic current into the substrate, activating a complex parasitic NPN transistor, which increases the power dissipation through its coupling with P1, P2, and N2 isolation n-wells. Functionality of other sensitive circuits on the chip is affected as well. As a result, in this operating mode, a substrate current is injected by the N1 switch through a lateral NPN transistor.

(b) REVERSE-FREEWHEELING. As shown in Fig. 2.7b, the LS switch N1 is always OFF, and the HS switch P1 is always ON, while P1 and N1 drive the inductive load. On the contrary to the prior forward-freewheeling configuration, during the P1/N1 dead time transition, the inductor current forces the P1 drain voltage to go above the V_{BAT} voltage level.

As a result, the performance of P1 is degraded, and a substrate current is injected into the substrate through the activation of the parasitic vertical PNP transistor. The gain (β) of such a parasitic device is technology-dependent. It ranges between 50 and 10 in standard HV-CMOS technologies, while it ranges between 1 and 0.001 in BCD technologies on high-doped substrates [26]. For high levels of injected currents, the substrate voltage can reach critical values which may forward bias other isolation junctions. In this operating mode, a substrate current is injected by P1 switch through a vertical PNP transistor.

(c) FORWARD-REVERSE. In this operating mode, the switch pairs P1 and N2 and P2 and N2 are driven on and off simultaneously as shown in Fig. 2.7c. During the dead time transition, the H-Bridge outputs are simultaneously driven belowground and above supply voltage levels. The NDMOS and DPMOS are at the same time injector and victim. Both the substrate PNP and NPN are activated, and the cumulated effects may trigger a parasitic PNPN thyristor structure and cause a latch-up.

These qualitative considerations can be considered as general guidelines to optimize the circuit floor planning and minimize substrate parasitic couplings. However, for an accurate evaluation of the circuit performance in the design phase, the effects of such parasitic bipolar devices must be quantified.

2.4.2 EMC Requirements for Automotive ICs

This section describes EMC requirements for automotive ICs, highlighting their impact on the generation of parasitic current into the substrate. Modern electronic systems for automobiles contain dozens of microprocessors and a variety of Radio frequency (RF) transmitters, receivers, and actuators. Such ICs have to pass very stringent tests to achieve the automotive certification in accordance with EMC standards.

In order to guarantee that a vehicle meets all the EMC requirements, it is firstly tested in a laboratory, and EMC issues are considered and tested at each stage of the electronic design process, starting from semiconductor ICs to electronic module levels (or Electronic control unit (ECU)) up to the final integration in a car [27]. The current 2004/104/European community (EC) directive provides EMC requirements and standardized tests between:

- car manufacturers and governments: radiation-based EMC compliance tests at the vehicle level in anechoic chambers [28];
- module suppliers and car manufacturers: EMC pre-compliance tests on electronic modules, also radiation based on both sides;
- module suppliers and IC suppliers: EMC pre-compliance radiation based on the module supplier side and conduction based on the chip supplier side.

In EMC, two domains are considered, Electromagnetic emission (EME) and Electromagnetic immunity (EMI) [29, 30]. EME defines the performance of a device with respect to its own emissions, while EMI defines the robustness of a device against interference with other devices. For semiconductor circuits, today full EMC simulations of complete chips are out of reach.

With some precautions, an incident EMC radiation on a circuit smaller than the radiation wavelength (IC size $\approx \frac{\lambda}{20}$) can be considered as a conducted signal to the IC input and power pins [31]. On the other hand, fast and large voltage and current changes at IC output and power pins are transformed into outward EMC radiation.

Based on these assumptions, IC designers can use SPICE for simulation of IC EMC and EMI against transients and RF disturbances [32]. In the frequency range up to several hundred MHz, IC coupling simulations can be carried out with lumped SPICE models. For instance, the higher-level EMC specifications are translated into simple pin-specific Direct power injection (DPI) specifications. Most relevant tests for IC-level specifications for SPICE simulations are standardized and specified in the following main documents:

- EMI:
  - ISO 7637-2 for transient pulses;
  - IEC 62132-4 for DPI [33];
- EME:
  - $1\,\Omega/150\,\Omega$ method specified in IEC 61967-4.

Tests are generally carried out on IC *global* pins. *Global* pins are defined as the IC pins that enter or leave the application module without any active component in between [27]. Supply pins or data bus pins are examples of global pins.

2.4.2.1 Transient Pulses

In vehicle electronic systems, the common battery supplies a number of modules (or ECU) as shown in Fig. 2.8. External radiated noise can affect each ECU through conduction and coupling via the power supply and data cables. Moreover, the switching of inductive loads such as motors or relays generates voltage surges which can affect the operation of electronic modules connected to the same power supply [34]. Often, power and data lines are bundled together creating additional capacitive and inductive coupling paths between the power line and data buses.

Electronic modules and ICs are subject to many transient signals generated by the vehicles electrical system. These signals are grouped and categorized according to their voltage amplitude and time duration in the international standard ISO 7637 parts 1 and 2. They are accurately reproduced during the test phase to distinguish whether modules withstand the disturbances and go back to normal operation afterward or if they are destroyed. Main transient signals that can occur in a vehicle power rail are shown in Fig. 2.9a. As easy to note, the voltage amplitude of the pulses is well beyond the expected battery operating range.

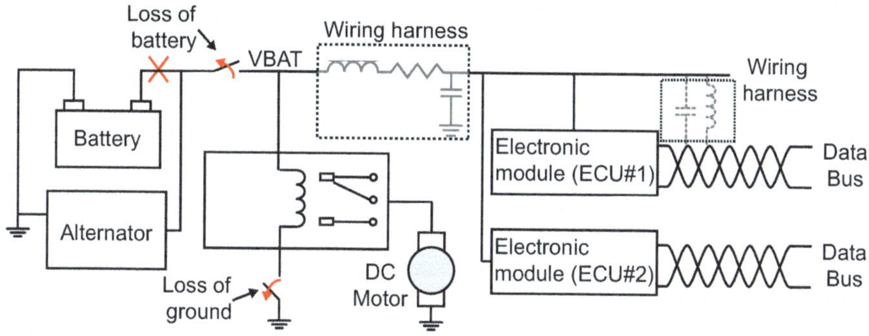

Fig. 2.8 Vehicle electronic system simplified architecture

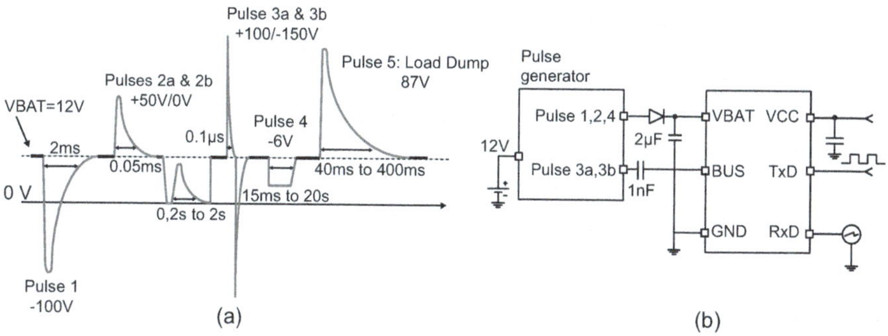

Fig. 2.9 (**a**) ISO 7637-2 main voltage surges in automotive systems and (**b**) circuit setup example for testing automotive transients

Initially applied to the ECU, today ISO pulses are applied directly to the ICs pins. Indeed, many semiconductor suppliers claim that ICs for automotive systems are designed to successfully pass alone standard ISO pulses. This requires the integration of internal protections that often require a larger die size and introduce many complexities within the already challenging design environment of HV ICs.

The setup for IC ISO pulse test is shown in Fig. 2.9b, where each pulse is directly applied to pins of the IC. The ISO pulses 1, 2, and 3 represent the disconnection of an inductive load in the wiring harness. If a discharged car battery is disconnected from the alternator while it is being charged, a large positive voltage spike is generated on the supply line. This is commonly referred to as *load dump* transient test and corresponds to pulse 5. Circuits that are directly connected to the supply line must withstand such spikes. In most cases, ISO 7637 test pulses are applied to IC pins connected to the 12 V battery via a reverse polarity diode in order to block any reverse current and to the data bus via a coupling capacitor (see Fig. 2.9b). Pulse 5, for its high energy, is rarely tested at IC level, and an external Zener clamp device is usually placed on each module for this purpose.

Since ISO pulse amplitudes are beyond supply voltage range, IC substrate BJT transistors are inevitably activated during such tests. In practice, the pulse generator has an internal resistance of 10 Ω, and a pulse with amplitude of up to ±100 V generates currents of up to ±10 A on a low impedance load. The IC must safely conduct this excess of current to avoid damages. A negative pulse of −100 V (ISO pulse1), applied either at the power supply or at the output node of an H-Bridge, would inevitably forward bias substrate junctions and generate a high substrate current.

Therefore, there is a need to track issues linked to the activation of substrate parasitic devices already during SPICE simulations of ISO pulses to prevent circuit damages and costly redesign loops.

2.4.2.2 Direct Power Injection

The EMI collected by cables connected to global pins of the IC is conveyed to the analog and digital cells causing operation failures. The most conclusive results of IC immunity against RF disturbances are achieved with a DPI test defined in the *IEC 62132-4* standard. This is the most widely used technique, especially in automotive applications where DPI signals can be applied directly to any IC global pin. In Fig. 2.10, the DPI setup for testing the CAN transceiver circuit is shown as example. The DPI disturbance is applied through an external low-pass filter to the circuit output pins CANH and CANL.

DPI testing is very simple and straightforward to perform [35]. The injected signal frequency ranges between 150 kHz and up to 1 GHz. For a forward power of 5 W, the EMI source with 50 Ω internal resistance reaches a maximum amplitude peak level of 45 V. Depending on the IC node impedance, the incident power is partially absorbed in the IC and causes disturbances in different ways. Disturbances can generate rectification and pumping in analog circuits due to component parasitics and nonlinearities [36]. Furthermore IC failures can be generated by the activation of substrate parasitic BJTs.

DPI testing can be easily simulated by chip designers using standard SPICE-based design tools up to several hundred of MHz. However, the estimation of DPI

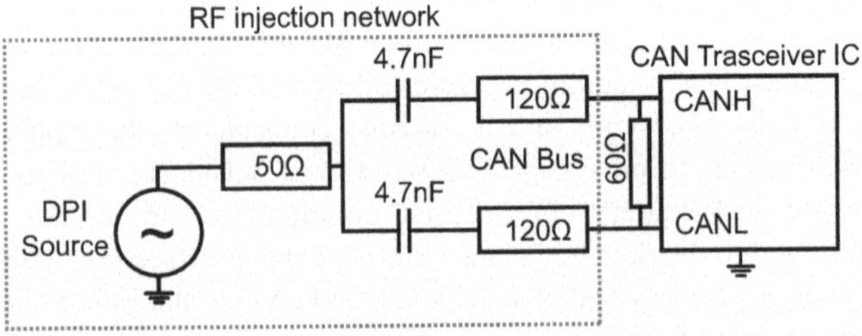

Fig. 2.10 Test setup for DPI test

effects requires long transient simulations over a long period of time with very large signals that exceed supply voltages. Indeed, the DPI compliance requires simulations for each DPI level for each frequency [37].

DPI simulation results are a first indication of the circuit's susceptibility to RF stresses [38]. However, effects due to substrate BJT devices are excluded during DPI simulations.

2.4.3 Failure Tests According to the ISO 16750-2

The ISO 16750-2 describes unexpected failures that can be caused by someone doing some maintenance on a vehicle. Such failures are often related to the activation of parasitic BJT transistors. The following tests are performed on automotive ICs to prevent these types of failures:

1. reverse battery voltage;
2. interruption of ground or battery lines;
3. extended short circuits.

Reverse Battery By wrongly reconnecting cables during car maintenance, it can happen that the polarity of the battery is applied in the reverse direction, thus resulting in damages on connected electronic modules. With a reverse applied voltage, a short circuit via diodes or transistors could occur in ICs, leading to electronic errors or permanent failures. Figure 2.11a shows a HS driver under reverse battery condition. At the module level, a series Shottky diode is placed in the positive supply line between the battery and ICs pins. By applying the battery voltage in the wrong polarity, the diode blocks the reverse currents, and the ICs are protected. To guarantee reverse battery protection even in the absence of the series diode, ICs integrate reverse battery solutions such as a back-to-back connected transistors or bulk switching techniques to avoid the activation of a parasitic vertical PNP transistor [39, 40].

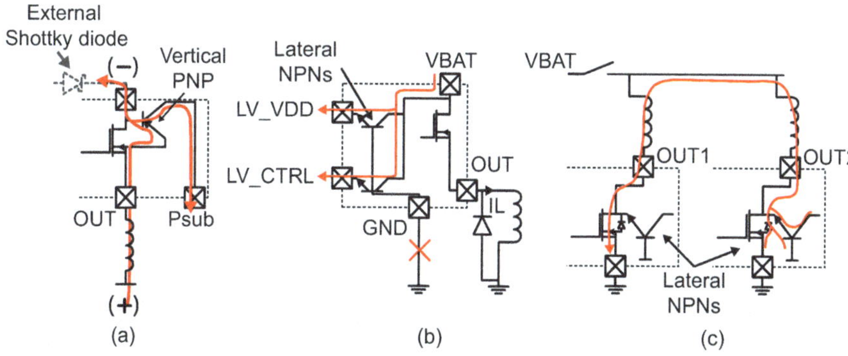

Fig. 2.11 Reversed battery condition (**a**), interruption of ground (**b**), and battery lines (**c**)

Ground or Battery Line Interruption In a HS driver, when the groundline is accidentally interrupted, the IC substrate remains floating. A substrate lateral NPN transistor may be activated between the VBAT supply (collector) and low-voltage pins (emitters). This example is illustrated in Fig. 2.11b. In many IC data sheets, it is recommended to use input resistors on low-voltage pins to limit the injection of current into the substrate due to the activations of parasitic BJT transistors. Similar effects are also caused by ground voltage offsets. The interruption of the power supply line VBAT may generate ground loops that make the current circulate between circuits connected to the same power line. When many LS drivers are connected to the same broken power supply (see Fig. 2.11c), the inductive current provided by the battery is interrupted and immediately circulates between the connected drivers. This can lead to the activation of a parasitic lateral NPN transistor in some of the LS drivers. Moreover, the reverse current is imposed by other LS channels and is potentially higher compared to the size of the exposed LS switch.

Extended Short-Circuit Protection for Low-Voltage Circuits Short circuit is a problem that can occur in a vehicle for IC global pins which carry low-voltage analog signals, such as the data bus pins, that can be accidentally shorted to the high-voltage battery. During the IC design, all cases of short-circuit conditions must be considered. Moreover, under these circumstances no reverse current is allowed. In this case the Controller area network (CAN) standard recommends that a transceiver must survive to bus wire short circuits within the wire, to the higher power supply and to the ground. Furthermore, most commercial circuits provide extended short-circuit protection to voltages in the −3 to 36 V range. Except for the duration of the short-circuit event, the electrical situation is similar to ISO pulses. These overvoltages can lead to the injection of a significant current into the substrate.

2.4.4 *High ESD Standards*

Specifications for ESD tests are detailed in the ISO 10605 standard. More specifically, automotive ICs might have I/O pins that withstand ±8 kV for contact discharge and ±15 kV for air discharge test methods [41].

An electrostatic discharge can trigger the parasitic substrate BJT transistors. Even if electrostatic discharges last a few tens of nanoseconds, when this happens, the entire chip substrate may be flooded with charge carriers. These charges generate a substrate current that can trigger a parasitic thyristor.

2.4.5 Summary

Power circuits, when driving an inductive load, generate self-induced substrate parasitic currents that must be monitored to ensure a proper operation of the circuit. Typical circuit topologies used for driving inductive loads when built in bulk technologies are susceptible to parasitic currents into the substrate. The analysis of circuit operating modes allows the identification of the substrate parasitic BJT transistors that are activated in Half- and Full H-Bridge circuits. This approach can be extended to other circuits driving an inductive load such as buck and boost switch mode power supplies.

Typically in automotive systems, with many other surrounding active elements, there are additional external disturbances, such as high-energy transient pulses and RF interferences, which can lead to the activation of substrate BJT transistors. The substrate devices which are activated during automotive EMC tests are highlighted in Table 2.2 [42].

The control of these parasitic devices poses additional challenges to the IC design. Indeed their effects are usually not addressed during the design and simulation phase of the circuit and are often cause of failures during the experimental evaluation of the first prototype. Only with a deep knowledge of the circuit architecture and of the magnitude of the substrate current can IC designers anticipate real effects.

2.5 Effects of the Substrate Current

The parasitic substrate current may cause adverse effects to HV ICs, such as soft failures (degradation of electrical properties) or hard failure (permanent damages).

A permanent damage can be caused by the triggering of a latch-up due to the cumulative effect of NPN and PNP parasitic transistors. Soft failures are caused by unexpected substrate couplings with other circuits and devices causing the degradation of some electrical properties. However, the malfunctioning of a circuit above a certain level and duration (latent failure) can also result in the permanent damage of the entire IC.

When a lateral NPN substrate transistor is activated, any other reverse-biased n-well becomes a collector of the parasitic lateral NPN transistor. The amount of current collected by each n-well depends on the distance and the separation area between collectors and the emitter. As a main effect, this generates extra power consumption in other circuits. Furthermore, n-wells are part of active and passive devices and correspond with nodes of analog and digital circuits. If a parasitic current is drawn from such nodes, circuit malfunctions are expected.

Similarly, when a vertical PNP transistor is activated, the substrate acts as a collector. The substrate is part of some active devices, and the substrate voltage shift affects their functionality. Global and local effects of the substrate current in ICs are provided in the following sections for different scenarios.

Table 2.2 List of automotive safety tests and related substrate devices

BCU pin	Test done at ECU module level on global pins	Test specifications at ECU module level with on board protection		Related test specifications at IC level		Risk of substrate injection		
			Typical external protection	Global pins	Local pins	Lateral NPN	Vertical PNP	Both
VBAT	Battery load dump	+85 V for 400 ms	+33 V TVS	VBAT 40 V				
VBAT	ISO7637 pulse 1	−100 V for 50 µs				X		
VBAT	ISO7637 pulse 2	+100 V for 2 ms					X	
VBAT	Reverse battery		Reverse battery diode	VBAT: −0.3 V to max. rating				X
VBAT	Outputs through load to VBAT: ISO7637 pulse 1	−100 V for 50 µs		−100 V for 50 µs through load min. 10 Ω		X		
VBAT	Outputs through load to VBAT: ISO7637 pulse 2	−100 V for 2 ms		−100 V for 2 ms through load min 10 Ω			X	
All	Inputs/outputs battery, ISO7637 pulse 3	±100 V (100 pF)						X
All	Inputs/outputs short circuit to GND with Vdrop->−2 V	2 ms		Short to −2 V		X (LS switch)		
All	Inputs/outputs short circuit to VBAT with Vdrop, VBAT+2 V			Short to VBAT +2 V			X (HS switch)	
All	EMC immunity	Bulk current injection 200 mA antenna radiated immunity		DPI 30 dBm	DPI 12 dBm			X
All	EMC emissions	Conducted and radiated emissions		CE IC Level				
All	ESD	±8 kV up to ±15 kV		±4 kV	±2 kV			X

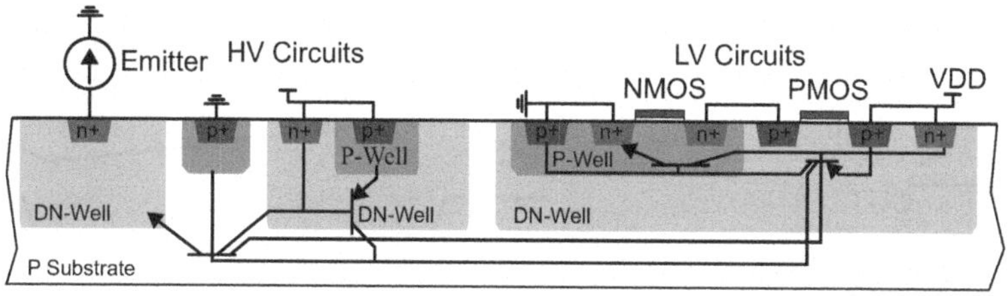

Fig. 2.12 HV technology cross section with latch-up equivalent circuit

2.5.1 *Global Effects*

2.5.1.1 Latch-Up

A latch-up occurs whenever a parasitic thyristor in the chip is activated, thus generating a low impedance path. The current flowing in this path, if not controlled, can cause permanent circuit damage. In CMOS logic, bipolar transistors which are part of latch-up loops are decoupled with special intra device guard rings to keep their loop gain from exceeding the unity. Moreover, the low impedance distribution of power and ground potentials over the entire chip enhances latch-up immunity.

Conversely, in a HV application, transient signals are no longer confined to power or logic I/O terminals. Substrate current generated by switching activities of power transistors may trigger a latch-up in internal regions of a CMOS device as well as at I/O power circuitry. Two parasitic PNPN paths are shown schematically in Fig. 2.12. For example, the activation of a lateral NPN transistor generates currents in any other n-well in the chip. An excessive voltage drop caused by the current collected in a n-well may forward bias a vertical PNP transistor, consequently triggering a thyristor either close to the output driver or in other regions such as in logic circuits.

2.5.1.2 Increase of Power Dissipation

The voltage and current waveforms of the lateral NPN transistor in a Half-Bridge switching period are shown in Fig. 2.13. In order to minimize the conduction of the substrate NPN, various techniques exist to control and optimize the dead time t_{dt} in the switch driver. A widely used technique, referred to as adaptive dead time control, uses a comparator to sense the OUT voltage. When the OUT node becomes negative, the synchronous LS switch is immediately turned on leading to a shorter diode conduction (i.e., synchronous rectification). However, in many designs, the dead time is fixed by the application and ranges between a few ns and hundreds of μs.

With reference to Fig. 2.13, both switches are off during the dead time, and the inductor current I_L flows through the DMOS switch body diode and the parasitic distributed NPN transistor:

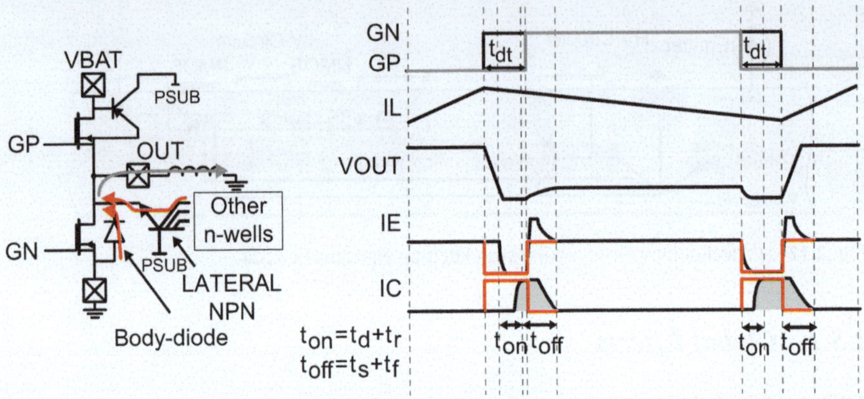

Fig. 2.13 NPN waveforms in a Half-Bridge switching period

$$I_L = I_D + I_{E_{\text{NPN}}} \tag{2.6}$$

The turn-on time t_{on} of the NPN transistor consists of an initial delay time, t_d, during which the base-emitter capacitance is charged. This delay is followed by the increase of the collector current quantified by the rise time t_r (see Fig. 2.13). Similarly, during the shutdown of the NPN device, the base current is reversed for the duration of time necessary to evacuate the excess of charge stored in the base-emitter junction. Typically, this duration is referred to as reverse recovery time, t_s. During this time the collector current continues to exist. Once the base-emitter charge is completely removed, the collector current decreases with a fall time t_f. General expressions and values for these timings can be derived in one-dimensional bipolar transistors under well-defined conditions [43]. However, these parameters are very difficult to determine for a three-dimensional substrate NPN transistor because of complex geometrical dependencies. They can only be determined thorough experimental study on dedicated test structures.

Here, to estimate power losses, the delays, rise/fall times, and reverse recovery times of NPN currents are neglected, and they are considered as square wave functions (see Fig. 2.13). The switching period is $\frac{1}{f_{sw}}$, and levels are 0 and $I_{E_{\text{NPN}}}$ for the emitter current, while 0 and I_{C_i} for the collector current with duty cycle $t_{dt}f_{sw}$ (t_{dt} is the dead time). Moreover the collector current at the i-th collector is

$$I_{C_i} = \alpha_i I_{E_{\text{NPN}}} \tag{2.7}$$

where α_i is the current gain of the lateral NPN between the NDMOS drain and the i-th collector.

These assumptions lead to inaccurate results, but they provide significant trends. Considering the NPN base voltage $V_{B_{\text{PNP}}} = V_{\text{PSUB}} = 0$, the average power dissipated by the substrate NPN is

$$P_{avg} \approx 2I_D \mid V_D \mid t_{dt}f_{sw} + 2I_{E_{\mathrm{NPN}}}t_{dt}f_{sw} \mid V_D \mid + 2I_{E_{\mathrm{NPN}}}t_{dt}f_{sw} \sum_{i=1}^{N}(\alpha_i V_{C_i}) \tag{2.8}$$

For a given reverse current, dead time losses are proportional to the dead time t_{dt} over the switching period ratio, to the gain α_i and to the collector voltage V_{C_i}. However, when no other n-wells are present in the IC layout $\alpha_i = 0$, or with all collector voltages are grounded $V_{C_i=0}$, then dead time losses are reduced to

$$P'_{avg} \approx 2I_D \mid V_D \mid t_{dt}f_{sw} + 2I_{E_{\mathrm{NPN}}}t_{dt}f_{sw} \mid V_D \mid \tag{2.9}$$

Equation (2.9) is commonly used to model dead time losses due to synchronous rectification.

However, in a real IC, there are many n-wells, and they are not all biased at 0 V. The extra power dissipation in the i-th n-well biased at a non-zero voltage is given by

$$\Delta P_{avg} \approx 2I_{E_{\mathrm{NPN}}}t_{dt}f_{sw} \sum_{i=1}^{N}(\alpha_i V_{C_i}) \tag{2.10}$$

The activation of a substrate lateral NPN leads to an increase of power dissipation. Minimization of the substrate current requires design and layout optimization to control, in addition to t_{dt}, $I_{E_{\mathrm{NPN}}}$, α_i, and V_{C_i} parameters as well. For this reason, in many designs, n-wells used as passive protections are grounded to minimize the power dissipation during the activation of the substrate lateral NPN transistor.

2.5.2 *Local Effects on Circuits and Devices*

The substrate current is generated by the activation of lateral NPN and vertical PNP parasitic transistors, which are created by various n- and p-type regions within the common substrate. However, such n-wells and p-wells which contribute to parasitics are also part of active and passive devices and correspond to nodes of analog and digital circuits.

N-wells are used to integrated low-voltage PMOS, isolated NMOS, and bipolar transistors. In HV technologies, n-wells are also used as drift regions for NDMOS transistors. P-wells are used in isolated NMOS, bipolar transistors, and drift regions of PDMOS devices. Moreover, some passive elements, such as resistors and capacitors, use n-well, p-well, and n/p diffusion regions as resistive material or to build reverse-junction capacitances.

When substrate parasitic devices are activated, each n-well or p-well in the circuit is sensitive to the substrate current. Therefore the parasitic substrate current sets certain limits on device usage in a circuit, since many are sensitive to minority carriers in the substrate. Depending on which substrate parasitic device is activated and the magnitude of parasitic current, the resulting effects are different.

2.5.2.1 Lateral NPN Effects

With the activation of a substrate lateral NPN, each n-region embedded in the chip substrate acts as parasitic collector. The effects on circuits vary depending on whether the collecting n-region is connected to a positive power supply or whether it is connected to a node of the circuit.

Impact on Low-Voltage PMOS Threshold Voltage If the n-region is an isolation n-well, connected to a positive power supply, the collected current can affect circuit function in two different ways. The primary effect is an increase in power dissipation as the collected current travels across the power supply. This effect was discussed in Sect. 2.5.1.2.

As a secondary effect, if a high current is collected by the isolating n-well, this current will cause a voltage drop in the n-well below the active devices. Figure 2.12 shows the case of a low-voltage PMOS transistor suffering from body effect because of the negative voltage shift of its local substrate.

$$V_t = V_{t0} + \gamma(\sqrt{\phi_s + V_{sb}} - \sqrt{\phi_s}) \tag{2.11}$$

This can cause unpredictable switching of logic circuits or affect the biasing point of PMOS devices.

Impact on Passive Devices Similarly, n-wells are used to fabricate large resistors. Resistor using n-regions should not be used, since the n-region itself is a collector of substrate current. Isolated resistors should use p-regions embedded in an n-region. However, to reduce nonlinearity effects due to depletion layer modulation, matched resistors are placed in separate isolation n-wells which are connected to one of the resistor terminals. If many resistors are connected in series, then substrate current will flows through the resistor chain. The same applies to capacitors.

Impact on Bipolar Devices In HV ICs, NPN and PNP bipolar devices are mostly used in precision circuits. Depending on the HV technology, there are different ways to build a bipolar transistor. In most cases both NPN and PNP transistors are based on lateral and vertical structures. However, each structure has an additional parasitic element with regard to the substrate. For example, the n-type base regions of a vertical PNP or the n-type collector of a NPN device are embedded directly into the substrate. Therefore, bipolar devices are highly susceptible to the substrate current.

Impact on NDMOS Devices A non-isolated DMOS transistor is also exposed to the substrate current. Besides leading to the activation of a lateral NPN, the NDMOS drain is also an efficient collector of substrate current. A NDMOS can be used as a current sensing device of a LS driver. When the LS driver generates a reverse current into the substrate ("belowground" condition), the NDMOS sensor collects part of the substrate current and can provide a wrong sensed current.

Impact on ESD Diodes The n-well belonging to an ESD protection structure based on a conventional dual diode configuration, or gate-grounded MOS or gate-coupled NMOS, is typically connected to input/output nodes of analog and digital circuits. An additional effect which can occur is that the ESD leakage increases under substrate current. If this leakage, for example, occurs at the inputs of a differential amplifier, then it is amplified by the gain of the differential pair. The substrate coupling with an ESD protection is discussed in the next Sect. 6.3.3.

Impact on Precision Analog Circuits In some circuit configurations, the isolation n-well is connected to device terminals for performance optimization of analog circuits. Thus, the n-region coincides with internal circuit nodes. Some analog circuit blocks use PMOS devices as differential pairs, cascode amplifiers, and current mirrors. The PMOS n-well (or body terminal) is, in certain cases, connected to the source to eliminate the body effect. Therefore, in the layout, PMOS transistors are placed in separated n-wells connected to the PMOS source terminal.

Although this approach increases the layout area, it minimizes degradation in analog circuit performance due to the body effect. On the other hand, these circuit configurations are exposed to the substrate current. Some or all the bias current may flow out of the cascode PMOS transistor through the n-well connected to its source. Likewise, the operating point of other circuits is shifted, which causes reduced accuracy, i.e., an increase of offset in a differential pair, or out-of-specification failures. Moreover, if the affected circuit is part of protection blocks, the malfunction of the protection circuit can cause permanent damage to the chip.

Circuits such as temperature sensors and bandgap references using PNP or NPN transistors [44] are very sensitive to the substrate current. The design of a parasitic-immune bandgap reference circuit is described in [45]. In the proposed bandgap circuit topology, NPN transistors having the collector connected to the power supply are adopted in order to reduce the impact of substrate couplings on circuit performance. With this approach, the substrate current collected by the NPN transistors flows through the supply pin and no internal node of the bandgap circuit is affected.

Many circuits include clamping Zener diodes based on vertical NPN structures. When such clamping structures are placed close to the power switch, the inherent PNP transistor can be activated by interrupting the current flow in the stack of Zener diodes. If this occurs, the overvoltage protection does not work properly, and the output switch may be damaged [46].

To conclude, in order to reduce circuit failures associated with the activation of a lateral NPN transistor, care should be taken in the connections of n-wells. In general, each n-well should be connected to a positive power supply rather than to the circuit's internal nodes. However, if the magnitude of substrate current is well known, reliable layout and design measures (i.e., a carefully designed IC floor plan or a circuit bias current higher than the leakage current) can be easily implemented.

2.5.2.2 Vertical PNP Effects

Substrates current are injected into the substrate when the p-body of a high-voltage transistor becomes forward biased. In this case a substrate vertical PNP transistor is activated with the substrate as a parasitic collector. The current injected into the substrate is related to the technology-dependent PNP current gain ($\alpha = I_C/I_E$).

Injected currents create positive voltage drops in the substrate that are proportional to the substrate doping. Then the substrate current flows out of the chip through substrate contacts. Excessive substrate voltage drop may cause the forward bias of n-wells to substrate junctions. Grounded n-wells used as passive protections in HV ICs belong to this category.

A forward-biased n-well to a substrate junction activates a lateral NPN transistor. If the lateral NPN transistor generates sufficient collector current to hold the PNP base current, then a sustained latch-up is triggered. However, if the parasitic transistor current gains are small, the conditions for triggering the latch-up are not satisfied (absence of self-sustaining positive feedback). This will only reinforce an increase of leakage currents in the circuit.

2.6 Conclusions

This chapter gives an overview of the basic HV ICs and technologies, substrate current origin, and effect on circuit's performance. Technology complexity and mask counts influence process costs. Maximizing the isolation between devices and minimizing costs involves a mutual trade-off. Broadly, most HV ICs are manufactured either with HV-CMOS or BCDs. However, HV-CMOS technology is particularly interesting due to the lower complexity and cost compared to BCD.

The origin and the magnitude of substrate current are specific to the circuit application, and no solution is available to quantify their effects. The use of isolation guards offers a fairly reliable solution to minimize substrate current effects. Such guards are typically placed around the sensitive or aggressor blocks following heuristic methods based mainly on the designer experience. This is done by extensive use of manual identification and analysis of potential problems, rather than an exhaustive analysis. Current systems are becoming increasingly complex, and a tool for in-depth analysis of the substrate current and a method to control them are therefore not only desirable but even essential.

The next chapters describe in detail the EPFL substrate model that includes parasitic substrate bipolar transistors with tools and methodologies developed to support the HV design flow.

References

1. R. Gariboldi, F. Pulvirenti, A 70 mΩ intelligent high side switch with full diagnostics. IEEE J. Solid State Circuits **31**(7), 915–923 (1996)
2. C. Contiero, P. Galbiati, M. Palmieri, G. Ricotti, R. Stella, Smart power approaches VLSI complexity, in *Proceedings of the 10th International Symposium on Power Semiconductor Devices and ICs, 1998. ISPSD 98* (1998), pp. 11–16
3. M.I.C. Simas, P. Santos, P. Casimiro, M. Lanca, Smart power in MOS technologies-an overview, in *Proceedings of the IEEE International Symposium on Industrial Electronics (ISIE)*, vol. 2 (July 1997), pp. 371–376
4. W. Pribyl, Integrated smart power circuits technology, design and application, in *Proceedings of the 22nd European Solid-State Circuits Conference (ESSCIRC)* (Sept 1996), pp. 19–26
5. B. Jayant Baliga, *Fundamentals of Power Semiconductor Devices* (Springer Science & Business Media, New York, 2010)
6. B.J. Baliga, An overview of smart power technology. IEEE Trans. Electron Devices **38**(7), 1568–1575 (1991)
7. B. Murari, F. Bertotti, G.A. Vignola, *Smart Power ICs: Technologies and Applications*, vol. 6 (Springer Science & Business Media, Berlin, 2002)
8. A. Elmoznine, J. Buxo, M. Bafleur, P. Rossel, The smart power high-side switch: description of a specific technology, its basic devices, and monitoring circuitries. IEEE Trans. Electron Devices **37**(4), 1154–1161 (1990)
9. S.L. Wong, S. Venkitasubrahmanian, M.J. Kim, J.C. Young, Design of a 60-V 10-A intelligent power switch using standard cells. IEEE J. Solid State Circuits **27**(3), 429–432 (1992)
10. H. Ballan, M. Declercq, M. Declercq, *High Voltage Devices and Circuits in Standard CMOS Technologies* (Springer, New York, 1999)
11. R.M. Forsyth, Technology and design of integrated circuits for up to 50 V applications, in *IEEE International Conference on Industrial Technology*, vol. 1 (IEEE, New York, 2003), pp. 7–13
12. A. Andreini, C. Contiero, P. Galbiati, A new integrated silicon gate technology combining bipolar linear, CMOS logic, and DMOS power parts. IEEE Trans. Electron Devices **33**(12), 2025–2030 (1986)
13. R. Roggero, G. Croce, P. Gattari, E. Castellana, A. Molfese, G. Marchesi, L. Atzeni, C. Buran, A. Paleari, G. Ballarin et al., BCD8sP: an advanced 0.16 μm technology platform with state of the art power devices, in *25th International Symposium on Power Semiconductor Devices and ICs (ISPSD)* (IEEE, New York, 2013), pp. 361–364
14. K.-S. Ko, S.-H. Lee, J.-W. Park, I.-W. Cho, K.-D. Yoo, J.-H. Kim, Proposal of 130 nm-60 V rated fully isolated LDNMOS with double Epi process, in *26th International Symposium on Power Semiconductor Devices & ICs* (IEEE, New York, 2014), pp. 398–401
15. A. Moscatelli, A. Merlini, G. Croce, P. Galbiati, C. Contiero, LDMOS implementation in a 0.35 μm BCD technology (BCD6), in *The 12th International Symposium on Power Semiconductor Devices and ICs* (IEEE, New York, 2000), pp. 323–326
16. V. Parthasarathy, R. Zhu, V. Khemka, T. Roggenbauer, A. Bose, P. Hui, P. Rodriquez, J. Nivison, D. Collins, Z. Wu, I. Puchades, M. Butner, A 0.25/spl mu/m CMOS based 70 V smart power technology with deep trench for high-voltage isolation, in *International Electron Devices Meeting, 2002. IEDM '02* (2002), pp. 459–462
17. P. Wessels, M. Swanenberg, H. van Zwol, B. Krabbenborg, H. Boezen, M. Berkhout, A. Grakist, Advanced BCD technology for automotive, audio and power applications. Solid State Electron. **51**(2), 195–211 (2007)
18. R. Minixhofer, N. Feilchenfeld, M. Knaipp, G. Röhrer, J.M. Park, M. Zierak, H. Enichlmair, M. Levy, B. Loeffler, D. Hershberger, F. Unterleitner, M. Gautsch, K. Chatty, Y. Shi, W. Posch, E. Seebacher, M. Schrems, J. Dunn, D. Harame, A 120 V 180 nm high voltage CMOS smart power technology for system-on-chip integration, in *22nd International Symposium on Power Semiconductor Devices & IC's (ISPSD)* (2010), pp. 75–78

19. S. Li, Y. Fu, Smart power technology and power semiconductor devices, in *3D TCAD Simulation for Semiconductor Processes, Devices and Optoelectronics* (Springer, Berlin, 2012), pp. 187–236
20. M. Stecher, N. Jensen, M. Denison, R. Rudolf, B. Strzalkoswi, M.N. Muenzer, L. Lorenz, Key technologies for system-integration in the automotive and industrial applications. IEEE Trans. Power Electron. **20**(3), 537–549 (2005)
21. J.P. Laine, L. Bertolini, M. Bafleur, C. Lochot, High-level substrate current effects in P-/-epitaxy/P+/-substrate smart power technologies, in *IEEE 15th International Symposium on Power Semiconductor Devices and ICs* (IEEE, New York, 2003), pp. 253–256
22. V. Parthasarathy, V. Khemka, R. Zhu, I. Puchades, T. Roggenbauer, M. Butner, P. Hui, P. Rodriquez, A. Bose, A multi trench analog + logic protection (M-TRAP) for substrate crosstalk prevention in a 0.25 μm smart power platform with 100 V high-side capability, in *Proceedings of the IEEE 16th International Symposium on Power Semiconductor Devices and ICs (ISPSD)* (May 2004), pp. 427–430
23. E.N. Stefanov, G. Charitat, N. Nolhier, P. Rossel, Transient behaviour of isolation architectures in smart power integrated circuits, in *European Conference on Power Electronics and Applications*, vol. 3 (1997), pp. 3–036
24. D.A. Grant, J. Gowar, *Power MOSFETs: Theory and Applications* (Wiley-Interscience, New York, 1989)
25. H. Casier, P. Moens, K. Appeltans, Technology considerations for automotive, in *Proceeding of the 34th European Solid-State Device Research Conference, 2004. ESSDERC 2004* (IEEE, New York, 2004), pp. 37–41
26. B. Murari, Power integrated circuits: problems, tradeoffs, and solutions. IEEE J. Solid State Circuits **13**(3), 307–319 (1978)
27. T. Steinecke, M. Bischoff, F. Brandl, C. Hermann, F. Klotz, F. Mueller, W. Pfaff, M. Unger, Generic IC EMC test specification, in *Asia-Pacific Symposium on Electromagnetic Compatibility (APEMC)* (IEEE, New York, 2012), pp. 5–8
28. M.A.K. Wiles, An overview of automotive EMC anechoic chambers, in *10th International Conference on Electromagnetic Interference & Compatibility, 2008. INCEMIC 2008* (IEEE, New York, 2008), pp. 75–80
29. H. Casier, Electronic circuits in an automotive environment. *Tutorial Presentation T3, ISSCC 2004, Feb. 15, 2004, San Francisco*
30. A.H.M. Van Roermund, H. Casier, M. Steyaert, *Analog Circuit Design* (Springer, New York, 2006)
31. O. Jović, Susceptibility of ICs to conducted electromagnetic interference. IEEE Trans. Electromagn. Compat. **34**, 123–137 (2009)
32. J.-M. Redouté, M. Steyaert, *EMC of Analog Integrated Circuits* (Springer Science & Business Media, New York, 2009)
33. S. Miropolsky, S. Frei, J. Frensch, Modeling of bulk current injection (BCI) setups for virtual automotive IC tests, in *EMC Europe* (2010)
34. E.B. Joffe, Power line transients on a bus due to the operation of the electrical systems, in *International Symposium on Electromagnetic Compatibility* (IEEE, New York, 1999), pp. 758–761
35. H. Pues, D. Pissoort, Design of IEC 62132-4 compliant DPI test Boards that work up to 2 GHz, in *International Symposium on Electromagnetic Compatibility (EMC EUROPE)* (IEEE, New York, 2012), pp. 1–4
36. J.-M. Redoute, M. Steyaert, An EMI resisting LIN driver in 0.35-micron high-voltage CMOS. IEEE J. Solid State Circuits **42**(7), 1574–1582 (2007)
37. A. Wieers, H. Casier, Methodology and case study for high immunity automotive design, in *Analog Circuit Design* (Springer, Berlin, 2006), pp. 219–237
38. F. Fiori, Susceptibility of smart power ICs to radio frequency interference. IEEE Trans. Power Electron. **29**(6), 2787–2797 (2014)
39. H.-P. Hong, J.-C. Wu, A reverse-voltage protection circuit for MOSFET power switches. IEEE J. Solid State Circuits **36**(1), 152–155 (2001)

40. S. Saponara, G. Pasetti, F. Tinfena, L. Fanucci, P. D'Abramo, HV-CMOS design and characterization of a smart rotor coil driver for automotive alternators. IEEE Trans. Ind. Electron. **60**(6), 2309–2317 (2013)
41. B. Deutschmann, F. Magrini, Y. Cao, Robustness of ESD protection structures against automotive transient disturbances, in *Asia-Pacific Symposium on Electromagnetic Compatibility (APEMC)* (IEEE, New York, 2010), pp. 1028–1031
42. W. Pfaff et al., *Generic IC EMC Test Specification* (ZVEI - German Electrical and Electronic Manufacturers Association, Electrical Components and Systems Division, Frankfurt am Main, 2017)
43. R.M. Warner, B.L. Grung, *Semiconductor Device Electronics* (Saunders College Publishing, Philadelphia, 1991)
44. G.C.M. Meijer, Integrated circuits and components for bandgap references and temperature transducers. PhD thesis, TU Delft, Delft University of Technology, 1982
45. W. Horn, H. Zitta, A robust smart power bandgap reference circuit for use in an automotive environment. IEEE J. Solid State Circuits **37**(7), 949–952 (2002)
46. M. Wendt, L. Thoma, B. Wicht, D. Schmitt-Landsiedel, A configurable high-side/low-side driver with fast and equalized switching delay. IEEE J. Solid State Circuits **43**(7), 1617–1625 (2008)

Chapter 3
Substrate Modeling with Parasitic Transistors

3.1 Mathematical Background and Meshing Concept

The drift-diffusion model for semiconductors for the stationary case is described by the system of elliptical Partial differential equations (PDEs) (Eq. (3.1)) where the unknowns are the electrostatic potential $V(x, y, z)$ (Poisson's equation) and the concentrations of electrons $n(x, y, z)$ and holes $p(x, y, z)$ (continuity equations):

$$\nabla \cdot (D_n \nabla n - \mu_n n \nabla V) = R(V, n, p) \tag{3.1a}$$

$$\nabla \cdot (D_p \nabla p + \mu_p p \nabla V) = R(V, n, p) \tag{3.1b}$$

$$-\lambda^2 \nabla^2 V = p - n + N \tag{3.1c}$$

where λ is the Debye length, N is the net doping concentration, and R is the recombination term where the Shockley-Read-Hall (SRH) model is considered.

Since no closed-form solution is available, several numerical methods have been developed to find a solution independently of electrical constraints, technology, or geometry. The Finite-difference method (FDM) and the Finite element method (FEM) are today implemented in modern TCAD software and used as reference from many engineers developing silicon technologies.

Anyhow, the physical system is divided in sub-domains where nonlinear equations can be well approximated. A mesh of geometrical points is then created and used to solve the set of equations (see Fig. 3.1). The key difference between the two methods comes from the meshing concept and the related approximations:

- FDM is based on a Cartesian meshing that may be uniform or not. From the mathematical point of view, the stability and convergence criteria are well-known, and the method is easily applied to simulate simple structures. However, for more complex applications, such a meshing is competitive with the FEM only if additional techniques are used, such as the Finite box method (FBM) [1].

P. Buccella et al., *Parasitic Substrate Coupling in High Voltage Integrated Circuits*, Analog Circuits and Signal Processing, https://doi.org/10.1007/978-3-319-74382-0_3

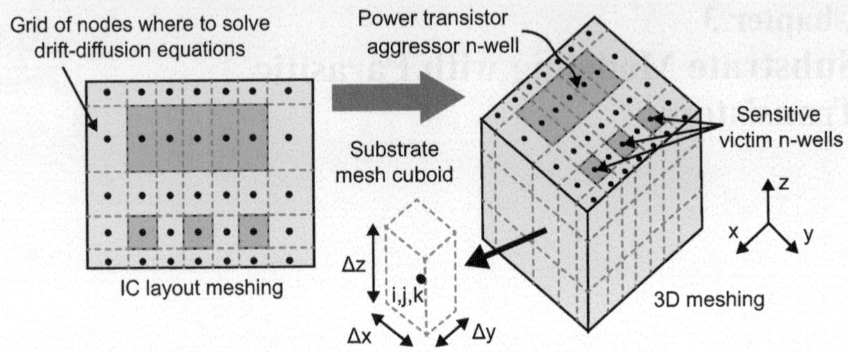

Fig. 3.1 HV IC layout with finite-difference discretization

- FEM is the best candidate for geometries that includes curvilinear shapes. Here the domain is divided in different volumes which can be non-orthogonal, e.g., the Delaunay triangulation. However, more computational effort is needed, and the convergence profile cannot be easily predicted.

IC layouts are composed by a set of masks that usually simplify the different geometries like implantation pockets with simple orthogonal polygons. In this case the FDM can be used to solve drift-diffusion equations. The entire substrate is then divided in rectangular cuboids where meshing points are connected to neighbor nodes. If the distance between the nodes is fixed, the mesh is *unifrom*; otherwise it is *nonuniform*. In the following section, only the orthogonal mesh is considered, i.e., the meshing cuboids are aligned to the Cartesian axes x, y, z. The indexes i, j, k will be used to order the points along x, y, z directions, respectively.

The nonuniform mesh is considered for the FDM scheme to Eq. (3.1). As shown in Fig. 3.2, a_i, b_j, and c_k are the distances between the i-th, j-th, and k-th meshing nodes, while $\Delta x_i, \Delta y_j, \Delta z_k$ are the dimensions of the substrate meshing cuboid referred to the node (i, j, k). The nodes indicated with $\pm 1/2$ are the ones at the boundaries between adjacent meshing cells (i.e., boundary nodes).

There is some arbitrariness in the definition of the meshing cuboid boundaries. In Fig. 3.2 different meshes are reported for the same node configuration (the example is in 2D but can be easily extended to 3D). Possible meshing strategies can be summarized as follows:

- **Node centered mesh (NCM):** the node (i, j) is midway between the left/right and top/bottom boundary nodes ($\overrightarrow{\Delta x_i} = \overleftarrow{\Delta x_i}$).
- **Line centered mesh (LCM):** the boundary nodes are exactly at the midpoints ($\overrightarrow{\Delta x_{i+1}} = \overleftarrow{\Delta x_i}$) between the different mesh nodes.
- **Uniform mesh (UM):** equal meshing spacing $a_i = \Delta x_i, b_j = \Delta y_j, c_k = \Delta z_k$ is assumed resulting in both NCM and LCM characteristics.
- **Nonuniform mesh (NUM):** the spacing between meshing nodes can be arbitrary, i.e., $\overrightarrow{\Delta x_{i+1}} \neq \overleftarrow{\Delta x_{i+1}} \neq \overleftarrow{\Delta x_i}$.

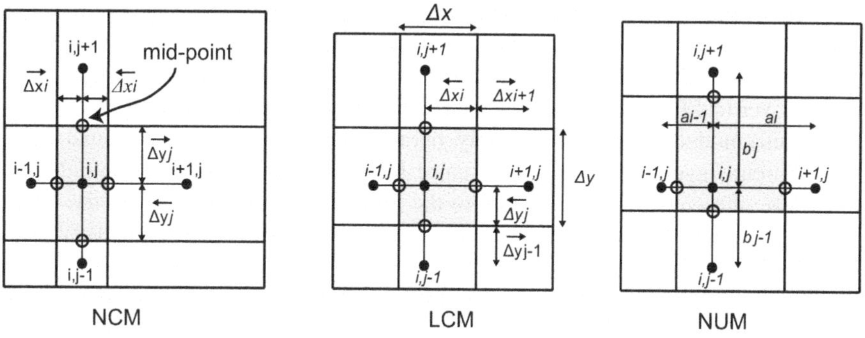

Fig. 3.2 Meshing schemes in 2D: NCM, LCM and completely NUM

Once defined the meshing concept, the FDM consists in linearizing the differential operators of Eq. (3.1) with a Taylor series around the point x_i,

$$f(x) = \sum_{n=0}^{\infty} \frac{(x - x_i)^n}{n!} \left. \frac{\partial^n f(x)}{\partial x^n} \right|_{x=x_i} \tag{3.2}$$

Considering the boundary nodes $i + \frac{1}{2}$ and $i - \frac{1}{2}$ distant, respectively, of $\overrightarrow{\Delta x}$ and $\overleftarrow{\Delta x}$ from i, the first derivative $f'(x_i)$ can be approximated using the central difference scheme, Eq. (3.3):

$$f'(x_i) = \frac{f(x_{i+\frac{1}{2}}) - f(x_{i-\frac{1}{2}})}{\Delta x} - \frac{\overrightarrow{\Delta x}^2 - \overleftarrow{\Delta x}^2}{2\Delta x} f''(x_i) + O(\Delta x) \tag{3.3}$$

When $\overrightarrow{\Delta x} = \overleftarrow{\Delta x}$, the accuracy is of the order of $O(\Delta x^2)$. Therefore, when the grid spacing tends to zero, the truncation error of the FDM vanishes as well. This gives rise to the concept of *meshing refinement*: the approximation of differential operators is more accurate when the number of meshing nodes increases, i.e., smaller distances for a given volume.

3.2 Modeling Methodology for Circuit Simulators

Today MOS transistors in circuits are simulated using compact models (e.g., BSIM, EKV, HSIM, etc.). These models may also include parasitic diodes and predict leakage currents. However, "unwanted" parasitic vertical and lateral BJTs exist between N- and P-doped wells in the substrate. Still, these are usually neglected since the corresponding PN junctions are normally reverse biased and should not affect reliability. However, if these junctions are forward-biased, a situation that may happen during power stage switching, some parasitic current paths may be activated and be responsible for severe reliability issues. Ideally, transistor models should also include substrate parasitic components to simulate these scenarios. For instance, vertical PNP BJTs could be characterized and be part of the parameterized cell (PCell) model [2, 3]. Unfortunately, parasitic lateral BJTs depend dramatically on the circuit layout (e.g., on the distance between transistors), and any attempt to simulate these effects will fail due to the high sensitivity to the geometry. Such geometry-dependent parasitic currents can still be simulated relying on a distributed model.

Instead of characterizing each combination of possible bipolar transistors between the different doped elements, a multi-junction approach should be preferred. As first proposed by Lo Conte et al. in [4] from Swiss Federal Institute of Technology of Lausanne (EPFL), the substrate can be partitioned to create an equivalent parasitic network of lumped components (Fig. 3.3). When a PN junction is detected, a diode is added to simulate injection or collection of charges; otherwise, a resistor is placed. To fill the gap in modeling parasitic lateral BJTs, specific components have additional terminals to generate and propagate minority carriers. In this way, even if the model does not include any BJT, interconnection between the different lumped components can predict the presence of parasitic transistors [5].

This substrate model is an extension of RC substrate noise models [6] that can simulate critical substrate parasitic currents as well. This involves three steps:

1. First, the entire substrate is according to the chip layout. The meshing process defines a three-dimensional grid of electrical nodes. It can be customized for any technology.
2. Next, the lumped components are instantiated between the nodes to create the equivalent parasitic network that will be simulated in circuit simulation tools.
3. Finally, the post-layout extracted netlist is simulated including the transistors and the substrate lumped components. The model parameters for the lumped

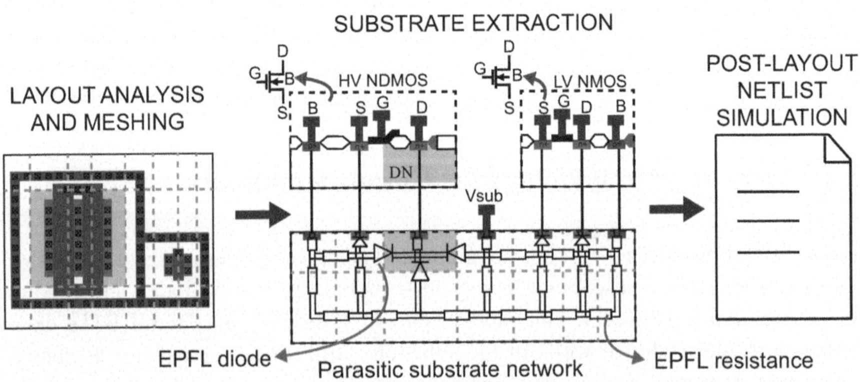

Fig. 3.3 Substrate model with distributed substrate parasitic network

components are extracted for a given technology process and stored in a model card.

The lumped components are then the core of the EPFL substrate model. They are derived applying the FDM approach to Eq. (3.1) in order to end with a system of algebraic equations that can be solved with well-known numerical methods. The spatial dependence is handled through the meshing of the circuit layout as in a distributed modeling approach. The number of equations to solve depends on the number of meshing nodes and is solved by circuit simulators. To this purpose, each equation must be "translated" in a circuit language, that is, current and voltages. These can be solved using Kirchhoff's voltage law (KVL) and Kirchhoff's current law (KCL). This is possible since a conservative system, such as the one satisfying the continuity equations, can be mapped onto equivalent potentials and fluxes [7]. New terminals are defined to "propagate" equivalent voltages V_{eq} and equivalent currents I_{eq} that are proportional to the minority carrier concentration and gradient, respectively.

The main drawback of distributed models is the number of equations and/or meshing points that must be introduced. The quantity that varies most rapidly in space is the charge density, and usually a meshing of the order of the Debye length is required (i.e., nanometers). Therefore, wherever an electric field builds up (e.g., PN junction interfaces and Metal-oxide semiconductor field effect transistor (MOSFET) channel region), the FDM grid used to solve the Poisson's equation requires a dense meshing. To minimize the simulation time, simulations are usually restricted to small volumes, i.e., small devices, as in TCAD software. To overcome this limitation, the following assumptions are used for the EPFL substrate model:

- **Quasi-neutrality hypothesis:** assuming that $\hat{n} \simeq \hat{p} \Rightarrow \nabla^2 V \simeq 0$, the Poisson's equation can be neglected in the system of PDEs (3.1), reducing the system complexity. This approximation can be assumed always valid and introduces small errors. In low-injection conditions, the electric field is negligible $E \simeq 0$, and no significant voltage drop appears. For high drift currents (high-injection regime), the difference $\hat{p} - \hat{n}$ is much smaller than the injected minority carrier concentration, and therefore it can still be neglected.
- **Minority carrier hypothesis:** the quasi-neutrality hypothesis implies that $\nabla\hat{n} \simeq \nabla\hat{p}$ applies and there is no need to solve the continuity equations for both holes and electrons but only for the minority carriers. Then, the continuity equation for the majority carriers can also be ignored. In conclusion the FDM is used only for one equation among the three required initially (3.1) which is a major simplification of the EPFL approach in regard to TCAD.

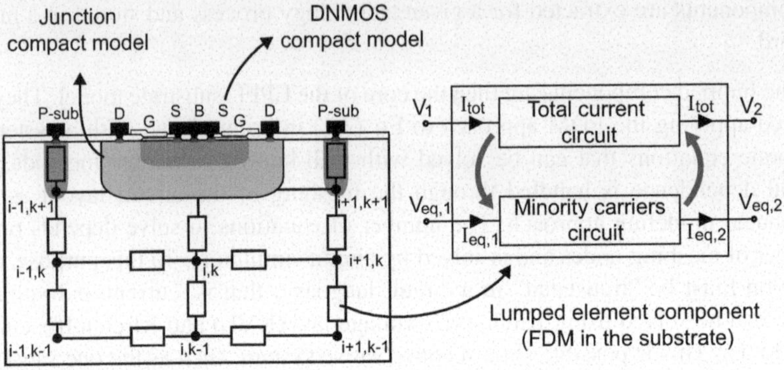

Fig. 3.4 Substrate modeling with lumped elements for circuit simulation. Each element is composed of a minority carrier circuit (MCC) coupled with a total current circuit (TCC)

Once discarded the Poisson's equation, it is still necessary to model regions where the electric field reaches some critical values. As a result, the complete substrate must rely on a semi-compact modeling approach: whereas compact models are used for transistors and junctions, the substrate needs for a distributed FDM model to solve the minority carriers transport equation (see Fig. 3.4). The variables in the continuity equation do not vary so much in space, and their computation requires few meshing points only.

The result of FDM applied to the minority carrier equation forms an algebraic system that is solved in circuit simulators by means of an equivalent circuit, called Minority carrier circuit (MCC). This circuit approach can be shown to be equivalent to the Transmission line model (TLM) method [8]. As a matter of fact, TLM is derived from analogies with electrical equations but is equivalent to FDM mathematical scheme. The MCC TLM circuit presented in the literature [9] has never been implemented to three-dimensional continuity equations and has always been restricted to a constant electric field (mean electric field). For instance, the Jankovic model [10] that uses TLM for BJT is only one-dimensional and cannot be extended to complex 3D multi-collector devices.

However, MCC gives only information on minority carriers, and post-processing is still needed to get more details on the parasitic currents and related voltage shifts. To have currents and voltages available, also a Total current circuit (TCC) must be simulated that takes into account the contributions from the majority carriers. These two equivalent circuits must be coupled and inserted between the substrate nodes.

3.3 Substrate Generalized Lumped Devices

The core of the EPFL substrate model consists in the generalized lumped devices shown in Fig. 3.5 which are merely an electrical representation of the discretized drift-diffusion equations that can be directly solved in the IC design flow. As already

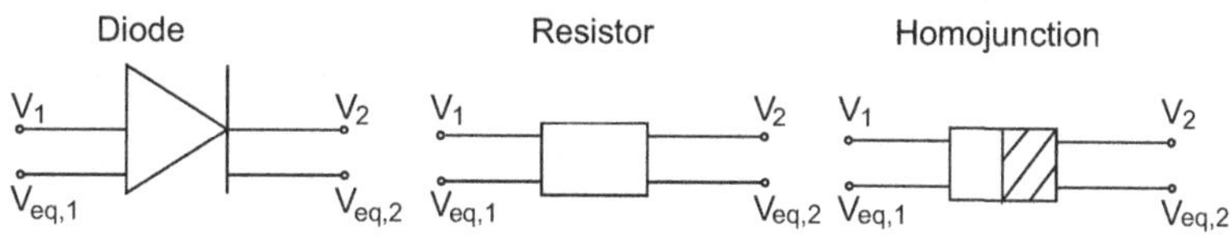

Fig. 3.5 Generalized lumped devices: EPFL diode, EPFL resistor, EPFL homojunction

mentioned, KVL and KCL are still used between the generalized devices to predict the distribution of currents, voltages, and minority carriers with standard circuit simulators.

Combining the MCC and the TCC, four-terminal devices are obtained. The resistor is the simplest element which propagates currents and voltages on $V_{1,2}$ and minority carriers on $V_{eq1,2}$. Besides the propagation, also the injection and the collection of minority carriers which occurs at PN junctions must be taken into account. A diode component that includes modulation of the depletion region is required. Finally, in all ICs, the doping profile is spatially dependent. In addition, there are regions with large doping discontinuities (PP+ or NN+) such as when encountering contacts, heavily p-type-doped substrate, or N+ buried layers. At these interfaces a model for homojunctions is required since the boundary conditions for the minority carriers are changed.

In the next subsections, the modeling of the substrate resistor, diode, and homojunction is reported with their equivalent circuits. The principal parameters used in the model are summarized in Appendix A, and they are still technology- and geometry-dependent.

3.3.1 EPFL Resistor

The resistance model can be derived from the current conservation law. In steady state the divergence of the total current density is null $\nabla \cdot J_{tot} = 0$. Applying the FDM to the partial derivatives, Eq. (3.4) is obtained where A_x, A_y, A_z are the areas normal to the x, y, z propagation directions [11].

$$\left(J_{x_{i+\frac{1}{2},j,k}} - J_{x_{i-\frac{1}{2},j,k}}\right) A_x + \left(J_{y_{i,j+\frac{1}{2},k}} - J_{y_{i,j-\frac{1}{2},k}}\right) A_y + \left(J_{z_{i,j,k+\frac{1}{2}}} - J_{z_{i,j,k-\frac{1}{2}}}\right) A_z = 0 \tag{3.4}$$

This result is equivalent to the KCL for the electrical node (i, j, k) where the sum of all currents entering in the finite volume $(\Delta x, \Delta y, \Delta z)$ must be zero. This is used to derive the "star configuration" circuit in Fig. 3.6 where the node (i, j, k) is connected to the boundary nodes of the meshing cell with six TCC [12].

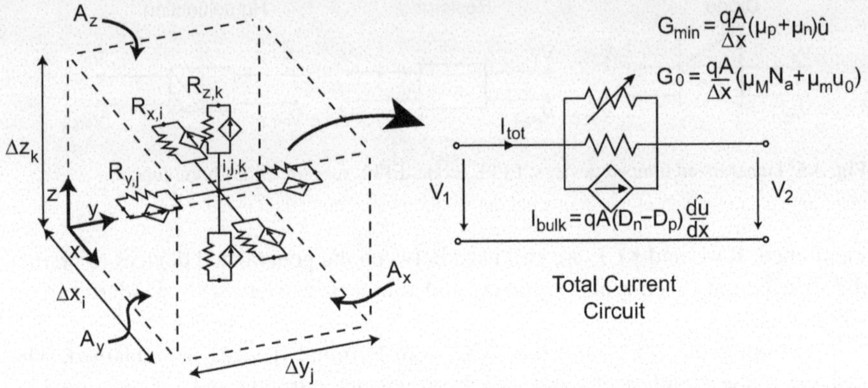

Fig. 3.6 Star configuration in the meshing volume for the total current circuit

The elements of the TCC sub-circuit are derived from the drift-diffusion current densities for holes and electrons:

$$J_n(x, y, z) = q\mu_n n(x, y, z)E(x, y, z) + qD_n\nabla n(x, y, z) \tag{3.5}$$

$$J_p(x, y, z) = q\mu_p p(x, y, z)E(x, y, z) - qD_p\nabla p(x, y, z) \tag{3.6}$$

To avoid writing twice the same equations, the generic carrier concentration $u(x, y, z)$ and a parameter $\alpha = 1$ for electrons and $\alpha = -1$ for holes are used. The subscript m denotes minority carriers, while M denotes majority carriers. Assuming that the diffusivity coefficients $D_{n,p}$ and the carrier mobilities $\mu_{m,M}$ are constant, the total current density at mid-nodes can be written for the x direction as:

$$J_{i+\frac{1}{2}} \simeq \underbrace{q[\mu_M(N + \hat{u}_{i+\frac{1}{2}}) + \mu_m(u_0 + \hat{u}_{i+\frac{1}{2}})]E_{i+\frac{1}{2}}}_{J_{ohm}} + \underbrace{q(D_n - D_p)\left.\frac{d\hat{u}}{dx}\right|_{i+\frac{1}{2}}}_{J_{bulk}} \tag{3.7}$$

The total current can be divided in two parts. The correction term J_{bulk} takes into account the difference between electron and hole diffusion currents. This is usually small, but still its impact can be non-negligible [13]. Due to the quasi-neutrality assumption, J_{bulk} is proportional to the gradient of injected minority carriers only and is implemented in the TCC as a controlled current source I_{bulk} coupled to the MCC. The ohmic contribution J_{ohm} is proportional to the electric field $E \simeq -\Delta V/\Delta x$ and so to the applied voltage drop. It is implemented as a resistive component between two nodes where the conductivity $\sigma(x, y, z)$ is modulated by the excess of minority carriers $\hat{u}(x, y, z)$:

$$\sigma(x, y, z) = q\left[\mu_M(N + \hat{u}(x, y, z)) + \mu_m(u_0 + \hat{u}(x, y, z))\right] \tag{3.8}$$

This general relation has two limiting cases:

- **Low injection** $\hat{u} \ll N$: the conductivity is determined by the doping concentration. This is the usual assumption in RC substrate models where the doping profile is used to set the substrate resistivity [6].

$$\sigma_0 = q(\mu_M N + \mu_m u_0) \simeq q\mu_M N \tag{3.9}$$

- **High injection** $\hat{u} \gg N$: the conductivity is modulated by the minority carriers excess concentration. In case of a p-type-doped substrate, J_{ohm} is mainly dominated by the drift current of electrons (minority carriers) since their mobility is higher than for holes.

$$\sigma_{HI}(x, y, z) \simeq q(\mu_p + \mu_n)\hat{u}(x, y, z) \tag{3.10}$$

Figure 3.6 reports the complete TCC (A represents the cross-section area) where the conductance G_0 is the standard substrate resistance and the conductance G_{min} is modulated by the minority carrier injection level.

Regarding the MCC, the drift-diffusion process in Eq. (3.7) requires the knowledge of the minority carrier concentration $\hat{u}$ and their gradient $\frac{d\hat{u}}{dx}$. This is obtained after solving the minority carrier continuity equation (3.11):

$$\nabla \cdot (\alpha D \nabla u - \mu u \nabla V) = \alpha R \tag{3.11}$$

The x, y, z components can be separated based on the linearity of the gradient operator ∇:

$$\frac{\partial}{\partial x}\left(\underbrace{\alpha q D \frac{\partial u}{\partial x} - q\mu u \frac{\partial V}{\partial x}}_{J_x}\right) + \frac{\partial}{\partial y}\left(\underbrace{\alpha q D \frac{\partial u}{\partial y} - q\mu u \frac{\partial V}{\partial y}}_{J_y}\right)$$
$$+\frac{\partial}{\partial z}\left(\underbrace{\alpha q D \frac{\partial u}{\partial z} - q\mu u \frac{\partial V}{\partial z}}_{J_z}\right) = \alpha q R(x, y, z) \tag{3.12}$$

This equation splits in three current density components along the x, y, and z directions. However, the general solution is not a linear combination of 1D solutions because the recombination term $R(x, y, z)$ is still a function of the volume. Applying the central difference scheme on the current divergence, the equation discretized inside a meshed space takes the form (3.13).

$$\frac{J_{x_{i+\frac{1}{2},j,k}} - J_{x_{i-\frac{1}{2},j,k}}}{\Delta x} + \frac{J_{y_{i,j+\frac{1}{2},k}} - J_{y_{i,j-\frac{1}{2},k}}}{\Delta y} + \frac{J_{z_{i,j,k+\frac{1}{2}}} - J_{z_{i,j,k-\frac{1}{2}}}}{\Delta z} = \alpha q R_{i,j,k} \tag{3.13}$$

It is now necessary to find a good approximation for the mid-interval quantities. In this case, J_x, J_y, J_z refer only to the minority carrier contributions; thus, they can be expressed for the x direction as in Eq. (3.14).

$$J_{i+\frac{1}{2}} = q\mu_{i+\frac{1}{2}} u_{i+\frac{1}{2}} E_{i+\frac{1}{2}} + \alpha q D_{i+\frac{1}{2}} \left.\frac{\partial u}{\partial x}\right|_{i+\frac{1}{2}} \tag{3.14}$$

As already assumed for the TCC, the central difference scheme along the path $x \in [x_i, x_{i+1}]$ is used once more for the electric field and the carrier's gradient:

$$E_{i+\frac{1}{2}} = -\left.\frac{\partial V}{\partial x}\right|_{i+\frac{1}{2}} = -\frac{V_{i+1,j,k} - V_{i,j,k}}{\overrightarrow{\Delta x_i} + \overleftarrow{\Delta x_{i+1}}} \quad \left.\frac{\partial u}{\partial x}\right|_{i+\frac{1}{2}} = \frac{u_{i+1,j,k} - u_{i,j,k}}{\overrightarrow{\Delta x_i} + \overleftarrow{\Delta x_{i+1}}} \tag{3.15}$$

Therefore, it is implicitly assumed that the diffusion current and the electric field have negligible second derivatives; thus, they are constant along the selected path. Using Eq. (3.15) into Eq. (3.14), the following expression is obtained (the same applies for the $-1/2$ point).

$$J_{x_{i+\frac{1}{2}}} = q\mu_{i+\frac{1}{2}} u_{i+\frac{1}{2}} \frac{V_i - V_{i+1}}{\overrightarrow{\Delta x_i} + \overleftarrow{\Delta x_{i+1}}} + \alpha q D_{i+\frac{1}{2}} \frac{u_{i+1} - u_i}{\overrightarrow{\Delta x_i} + \overleftarrow{\Delta x_{i+1}}} \tag{3.16}$$

Assuming Einstein's relation applies for D and assuming that the mobility is constant, only an estimation of the midpoint minority carrier density $u_{i+\frac{1}{2}}$ is missing. Now, if the diffusion current is assumed to be piecewise constant, the spatial variation of the carrier concentration can be interpolated by a linear function:

$$u_{i+\frac{1}{2}} = \frac{u_i \overrightarrow{\Delta x_i} + u_{i+1} \overleftarrow{\Delta x_{i+1}}}{\overrightarrow{\Delta x_i} + \overleftarrow{\Delta x_{i+1}}} = \frac{u_i R_{g,i} + u_{i+1}}{1 + R_{g,i}} \tag{3.17}$$

In Eq. (3.17) the geometrical ratio parameters R_g are defined as

$$R_{g,i} = \frac{\overrightarrow{\Delta x_i}}{\overleftarrow{\Delta x_{i+1}}} \tag{3.18}$$

Note that $R_g = 1$ only for LCM where the mid value is considered; otherwise, the quantities $u_{i+\frac{1}{2}}, u_{j+\frac{1}{2}}, u_{k+\frac{1}{2}}$ are meshing-dependent. Substituting Eq. (3.16) into Eq. (3.13), the final discretized continuity equation (3.19) is obtained.

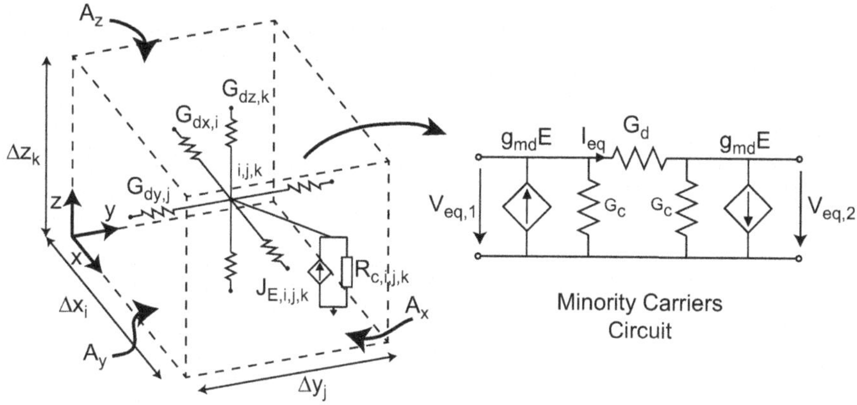

Fig. 3.7 Star configuration in the meshing volume for the minority carrier circuit

$$
\begin{aligned}
&\left(qD_{i+\frac{1}{2}}\frac{u_{i+1}-u_i}{\overrightarrow{\Delta x_i}+\overleftarrow{\Delta x_{i+1}}}-qD_{i-\frac{1}{2}}\frac{u_i-u_{i-1}}{\overrightarrow{\Delta x_{i-1}}+\overleftarrow{\Delta x_i}}\right)A_x\\
&+\alpha q\left(\mu_{i+\frac{1}{2}}\frac{u_{i+1}+u_iR_{g,i}}{1+R_{g,i}}E_{x_{i+\frac{1}{2}}}-\mu_{i-\frac{1}{2}}\frac{u_{i-1}+u_iR_{g,i}}{1+R_{g,i}}E_{x_{i-\frac{1}{2}}}\right)A_x\\
&=qR_{i,j,k}\Delta x\Delta y\Delta z
\end{aligned}
\tag{3.19}
$$

Extending this result to 3D, six similar expressions with different orientations are obtained, plus a recombination component depending on the volume of the meshing cells. This can be translated in a star circuit for the node (i, j, k) with an additional element for the recombination term (see Fig. 3.7). Fortunately, the volume can be expressed as in Eq. (3.20) allowing to split the recombination term as a parallel contribution along the different branches. The division factor is called dimensional parameter and is $d_m = 3$ in 3D.

$$
\Delta x\Delta y\Delta z=\frac{\overrightarrow{\Delta x}+\overleftarrow{\Delta x}}{3}A_x+\frac{\overrightarrow{\Delta y}+\overleftarrow{\Delta y}}{3}A_y+\frac{\overrightarrow{\Delta z}+\overleftarrow{\Delta z}}{3}A_z
\tag{3.20}
$$

In conclusion the linearized equation (3.19) can be expressed in terms of the equivalent MCC circuit presented in Fig. 3.7. The electrical analogy relies on the definition of the equivalent voltage proportional to the excess concentration of the minority carriers $V_{eq} = q\hat{u}$, where the elementary charge q is added for scaling purposes. Three components similar to the Linvill's circuit ones can be recognized [14]:

$$
G_d=\frac{AD_m}{\Delta x}\quad G_c=\frac{A\Delta x}{2\tau_m}\quad g_{md}=A\mu_m\frac{V_{eq,1}+V_{eq,2}R_g}{1+R_g}
\tag{3.21}
$$

1. **The diffusance** G_d is a conductance-dependent on the diffusivity D_m and regulates the diffusion current I_{eq} which is proportional to the minority carrier gradient $\frac{d\hat{u}}{dx}$.
2. **The combinance** G_c is a conductance-dependent on the minority carrier lifetime (assuming the SRH model, Eq. (3.22)) and regulates the fraction of minority carriers that recombine in the considered volume.

$$R_{\text{SRH}} = \frac{\hat{u}(N + u_0 + \hat{u})}{(\tau_m N + \tau_M u_0) + (\tau_m + \tau_M)(\hat{u} + n_i)} \simeq \frac{\hat{u}}{\tau_m} \tag{3.22}$$

3. **The driftance** g_{md} is a transconductance that weights the contribution of the electric field (drift term) and is dependent on the average minority carrier concentration.

The computation of V_{eq}, I_{eq} allows to compute G_{min} and I_{bulk} of Fig. 3.6 introducing couplings between the two circuits ($E \simeq 0$), MCC and TCC. For low-injection condition ($E \simeq 0$), MCC and TCC are decoupled, and the resulting network is equivalent to the substrate modeling methodology presented in [15], while in high injection the circuits must be solved with coupling. For the TCC, currents or voltages can be imposed as for real components, while for the MCC it is assumed that all the minority carriers recombine at ohmic contacts, meaning that $V_{eq} = 0$ (Dirichlet's condition) .

The equivalent circuit model for the EPFL substrate resistor is summarized in Fig. 3.8 for the P- and N-doped case where the TCC (Fig. 3.6) and MCC (Fig. 3.7) sub-circuits are coupled.

3.3.2 *EPFL Diode*

The generalized model of the EPFL diode is a combination of a P and N resistors plus a model for the junction region. Since in a PN junction there is an exchange

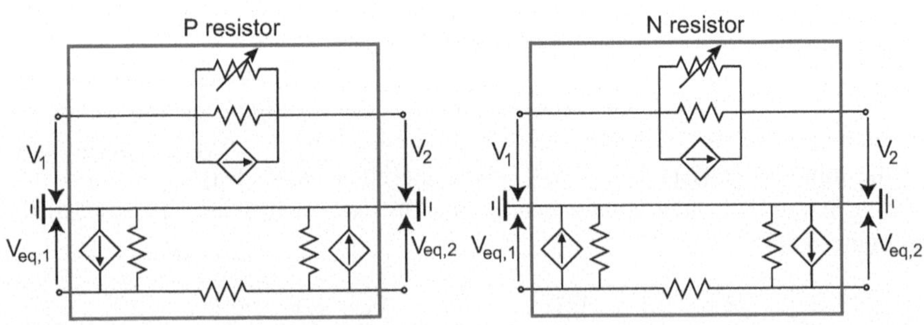

Fig. 3.8 Equivalent circuit of EPFL resistor

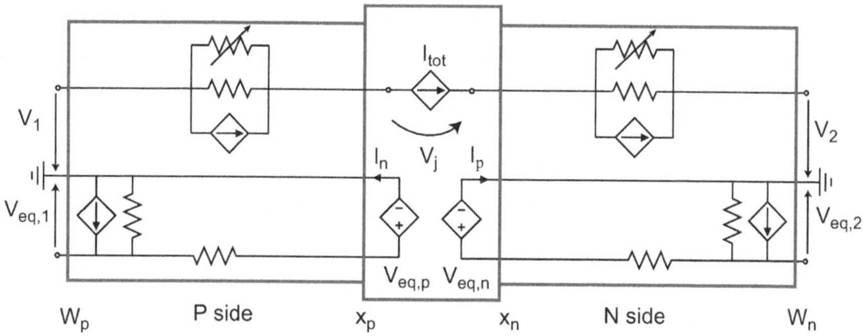

Fig. 3.9 Equivalent circuit of EPFL diode

of majority and minority carriers, the MCC sub-circuits cannot be interconnected since one refers to holes while the other to electrons. To model minority carriers at the boundaries of the depletion region (x_p in the P-side and x_n in the N-side), the junction must be modeled as nonlinear voltage sources.

Consider V_j as the voltage drop across the depletion region. In the low carrier injection, Shockley's boundary conditions can be used leading to the well-known $\simeq e^{V_j/V_t}$ voltage dependence of the minority carrier density. In high injection, the Misawa boundary conditions apply since a $\simeq e^{V_j/2V_t}$ dependence is expected at high-bias voltages [16]. As shown in Fig. 3.9, the general boundary conditions equation (3.23) are modeled as voltage sources (used in place of equivalent minority carrier concentrations) controlled by the potential V_j.

$$V_{eq,n} = q\hat{p}_n \simeq q\frac{N_d}{2}\left(\sqrt{1+\frac{4n_i^2\left(e^{\frac{V_j+V_r}{V_t}}-1\right)}{N_d^2}}-1\right) \tag{3.23a}$$

$$V_{eq,p} = q\hat{n}_p \simeq q\frac{N_a}{2}\left(\sqrt{1+\frac{4n_i^2\left(e^{\frac{V_j+V_r}{V_t}}-1\right)}{N_a^2}}-1\right) \tag{3.23b}$$

In these expressions n_i is the intrinsic carrier concentration of silicon, $\hat{n}_p, \hat{p}_n$ are the excess carrier concentrations at the boundaries of the depletion region, N_a and N_d the doping concentration of P- and N-side, respectively, and V_t the thermal voltage. A voltage correction term V_r is introduced since the Fermi level splitting must be corrected for the enhanced density of charges. From the definition of Fermi level, V_r is dependent on the excess in minority carriers as follows:

$$V_r = \underbrace{V_t \ln\left(\frac{N_a + \hat{n}_p}{N_a + \hat{n}_1}\right)}_{p\text{-side}} + \underbrace{V_t \ln\left(\frac{N_d + \hat{p}_n}{N_d + \hat{p}_2}\right)}_{n\text{-side}} \tag{3.24}$$

In Eq. (3.24) $\hat{n}_1$ is the excess concentration of electrons at the geometrical boundary W_p of the P-side and $\hat{p}_2$ the excess concentration of holes at the geometrical boundaries W_n of the N-side.

For the TCC sub-circuit, a current source I_{tot} in the depletion region is added to the model and coupled to the minority carrier circuit. Equation (3.25) shows that this current is composed of electron's current at the P-side boundary x_p and of hole's current at the N-side boundary x_n. Both components include diffusion and drift terms: the first being relevant for low-injection and the latter for high-injection regimes. Additionally, the Sah-Noyce-Shockley (SNS) correction term I_{SNS} is included to model leakage currents at small forward bias voltages [17].

$$I_{tot} = \underbrace{I_{n,drift}(x_p) + I_{n,diff}(x_p)}_{I_n} + \underbrace{I_{p,drift}(x_n) + I_{p,diff}(x_n)}_{I_p} + I_{\text{SNS}} \tag{3.25}$$

Taking into account all these aspects, the equivalent circuit of the EPFL diode is reported in Fig. 3.9.

3.3.3 *EPFL Homojunction*

Two regions of the same type, but differing from the doping density, constitute a homojunction. This is used in integrated circuits to create ohmic contacts with highly doped P+ or N+ non-rectifying homojunctions [18]. At the low-/high-doping transition, a space-charge region is created and a built-in potential appears [19]. For simplicity such a space-charge region is neglected in the model, but still, the homojunction modifies the boundary conditions for the minority carriers, which, when neglected, may lead to large errors when simulating parasitic BJTs.

From the point of view of the total current, a PP+ or NN+ homojunction is equivalent to a discontinuity in the resistivity. Connecting together two TCCs with different doping seems reasonable, as reported in the equivalent circuit of Fig. 3.10. But on the other hand, the boundary conditions used to solve the continuity equations for the minority carriers change dramatically. To avoid errors in computing the diffusion current, the high-to-low-doped homojunction interface must be properly modeled.

The equilibrium band diagram of a diode is reported in Fig. 3.11 with P+ and N+ contacts at each end. Barriers for electrons in the P-side, and for holes in the N-side, appear at the NN+ and PP+ junctions and disturb the diffusion of carriers to the contacts. By extension the same happens each time a doping discontinuity is present. Consequently, across homojunctions, minority carriers accumulate in the

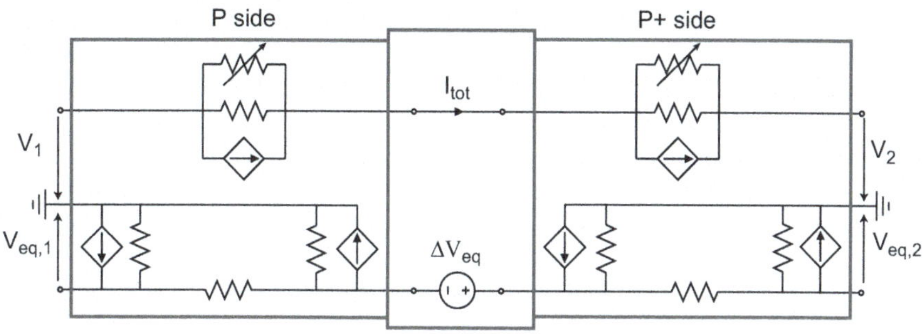

Fig. 3.10 Equivalent circuit of EPFL homojunction

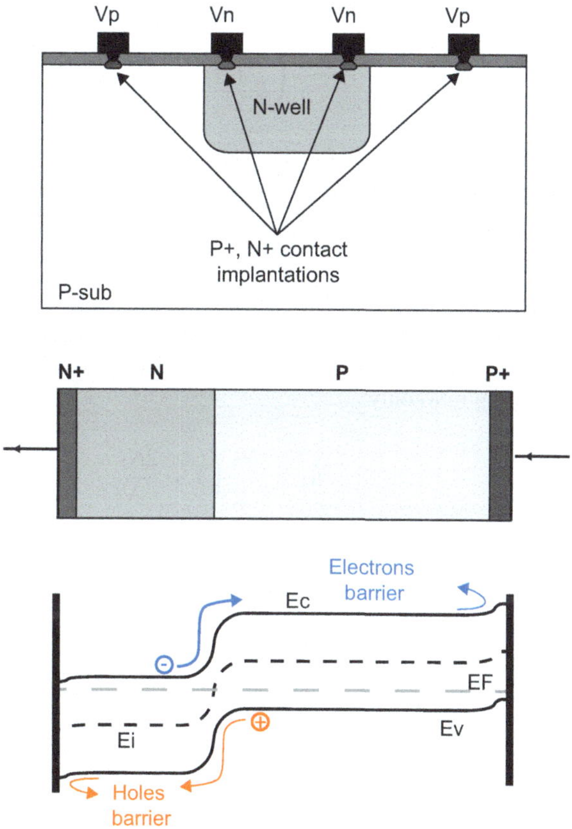

Fig. 3.11 Diode cross section with high-doped end contacts and relative band diagram. E_C: conduction band; E_V: valence band; E_F: Fermi level; E_i: intrinsic Fermi level

low-doped side and deplete the high-doped side. This discontinuity in the carrier concentration comes from the discontinuity in the fixed charge distribution.

To fulfill Boltzmann's statistics slightly out of equilibrium, the law of action mass imposes a relation between the minority carriers injected $\hat{u}_1$ on the P-/N-side and $\hat{u}_2$ on the P+/N+ side. Equation (3.26) expresses the relationship between the minority carrier concentration at the two homojunctions.

$$(N_1 + \hat{u}_1)(u_{01} + \hat{u}_1) = (N_2 + \hat{u}_2)(u_{02} + \hat{u}_2) \tag{3.26}$$

This condition can be modeled as discontinuity of minority carrier concentration and then as a drop in the equivalent voltages $\Delta V_{eq} = q(\hat{u}_2 - \hat{u}_1)$ in the MCC.

To determine the voltage discontinuity, the excess concentrations of minority carriers $\hat{u}_{1,2}$ must fulfill the following equation derived from Eq. (3.26) (at equilibrium $Nu_0 = n_i^2$):

$$\hat{u}_1^2 + (N_1 + u_{01})\hat{u}_1 - (N_2 + u_{02})\hat{u}_2 - \hat{u}_2^2 = 0 \tag{3.27}$$

Assuming that the index 1 is for the low-doped side and that the index 2 is for the high-doped one (i.e., $N_2 > N_1$), the solution is:

$$\hat{u}_2 = \frac{N_2 + u_{02}}{2}\left(\sqrt{1 + \frac{4(N_1 + u_{01} + \hat{u}_1)\hat{u_1}}{(N_2 + u_{02})^2}} - 1\right) \tag{3.28}$$

The equivalent voltage discontinuity ΔV_{eq} has to fulfill Eq. (3.28), which can be interpreted as a reflection of minority carriers at the PP+ or NN+ interfaces that change the boundary conditions for recombination. Two asymptotic cases can be analyzed:

- **Low-injection condition:** when the minority carrier density injected is much less than the doping density (i.e., $\hat{u}_1 \ll N_1$), the square root can be approximated as $\sqrt{1+x} \approx 1 + x/2$, and the minority carrier discontinuity is dependent on the doping ratio only:

$$\hat{u}_2 \simeq \frac{N_2}{2}\left(1 + \frac{2N_1\hat{u}_1}{N_2^2} - 1\right) = \frac{N_1}{N_2}\hat{u}_1 \tag{3.29}$$

- **High-injection condition:** when the injected minority carrier density is much higher than the doping density (i.e., $\hat{u}_1 \gg N_1$), the minority carrier discontinuity tends to vanish due to the lowering in the barrier potential:

$$\hat{u}_2 \simeq \frac{N_2}{2}\left(\frac{2\hat{u}_1}{N_2} - 1\right) \simeq \hat{u}_1 \tag{3.30}$$

The minority carrier concentration discontinuity ΔV_{eq} induced by the discontinuity in doping depends on the applied voltage (i.e., the injection level). Numerical simulations of a PN junction have been performed with TCAD to investigate the voltage-dependent barrier potentials on minority carriers at PP+, NN+ interfaces [20].

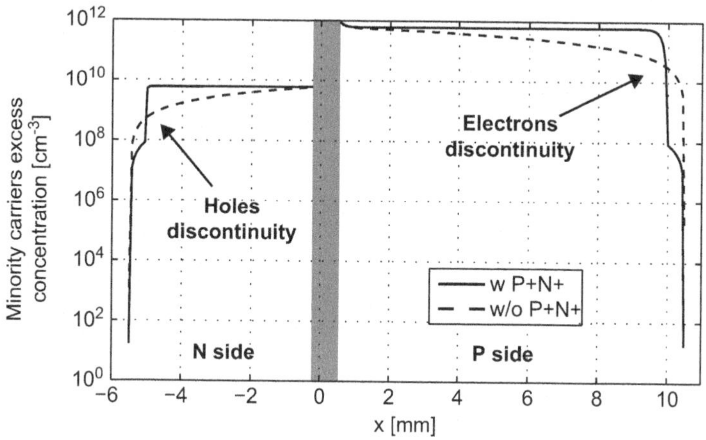

Fig. 3.12 Simulated minority carrier discontinuity from TCAD with and without P+ and N+ contacts homojunctions at 400 mV forward bias

Figure 3.12 plots the discontinuity for the excess carrier concentrations for a forward-biased n-well in a low-doped substrate with N+ and P+ contacts. Compared to the same PN junction without P+ and N+ highly doped layers, the boundary conditions for the minority carriers change considerably. As expected from Eq. (3.28), simulations reveal a drop of two decades on the N-side ($N_N/N_{N+} = 10^2$) and four decades on the P-side ($N_P/N_{P+} = 10^4$) for hole and electron concentrations, respectively.

This means that the diffusion current is also reduced as it scales with the gradient of the minority carriers concentration. Figure 3.13 reports the forward bias IV characteristics of the diode when the doping concentration at the P+ contact is varied from 10^{15} to 10^{20} cm^{-3}. Two different behaviors are evidenced in low- ($V < 1$ V) and high-injection regimes ($V > 1$ V):

- **Low-injection case (effects on diffusion currents):** diffusion currents are reduced with highly doped contacts proportionally to the ratio in the doping discontinuity. As shown in Fig. 3.13, the P+ contact width has no effect on the discontinuity of the electron concentration at PP+ interface, i.e., it has negligible impact on the diffusion current.
- **High-injection case (effects on drift currents):** for high forward bias, the current is increased with highly doped contacts. This comes from a change in the transport process that shifts from diffusion to drift. As expected, the resistance is lowered by the large injection of minority carriers (conductivity modulation). Moreover, for high applied voltages, the electric field accumulates carriers on the low-high interfaces and, in some cases, reverses the gradient in the carrier concentration.

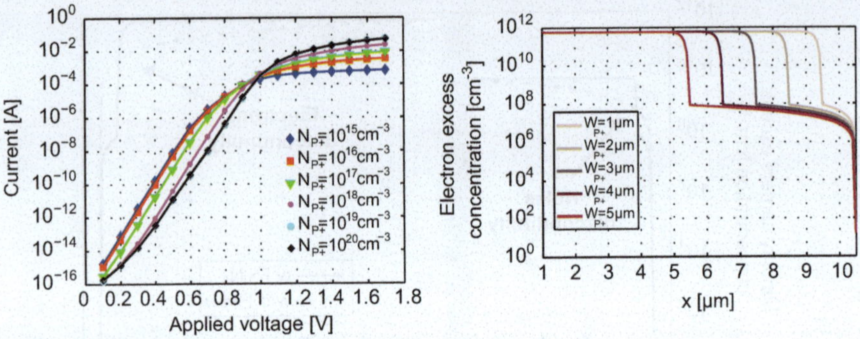

Fig. 3.13 (Left) Simulated diode IV curves for different doping concentrations of P+ contact region. (Right) Electrons discontinuity variation as a function of the different P+ width for $V = 400\,\text{mV}$

3.4 Extensions of the Model

In the previous sections, only the steady-state condition was analyzed. However, this is not complete to simulate real cases. In the following paragraph, three generalizations are carried out for the EPFL substrate model to be used in practical IC design. The first one is an extension of the FDM approach to model non-steady-state operation and will include capacitive coupling in the equivalent circuits of GLD. The second extension will deal with the impact of specific CMOS technology solutions on the model. Finally the last development will include junction breakdown mechanisms.

3.4.1 AC and Transient Model

In order to include transient effects in the EPFL substrate model, the nonstationary drift-diffusion continuity equation must be considered. In one dimension, the continuity equation for electrons $\hat{n}(x)$ is given by (similar equation applies for holes)

$$\frac{\partial}{\partial x}\left(\underbrace{q\mu\hat{n}(x,t)E(x) + qD\frac{\partial \hat{n}(x,t)}{\partial x}}_{J_{nx}}\right) = qR(x,t) + q\frac{\partial \hat{n}(x,t)}{\partial t} \tag{3.31}$$

The last term holds for time-dependent charge effects. Applying the FDM to the spatial derivative, Eq. (3.31) is linearized between N discretized points. The node i is separated by Δx_{i-1} from the "preceding" $i-1$ and by Δx_i from the "following" node $i+1$. Using the central difference scheme for the minority carrier current density, the continuity equation satisfies

$$\frac{\partial J_{nx_i}}{\partial x} \simeq \frac{J_{nx_{i+\frac{1}{2}}} - J_{nx_{i-\frac{1}{2}}}}{\frac{\Delta x_i + \Delta x_{i-1}}{2}} = q\left(R_i + \frac{\partial \hat{n}_i}{\partial t}\right) \tag{3.32}$$

As discussed in Sect. 3.3.1, the midpoint currents are approximated with the following form:

$$J_{nx_{i+\frac{1}{2}}} = q\mu_{i+\frac{1}{2}}\hat{n}_{i+\frac{1}{2}}\frac{\hat{V}_i - \hat{V}_{i+1}}{\Delta x_i} + qD_{i+\frac{1}{2}}\frac{\hat{n}_{i+1} - \hat{n}_i}{\Delta x_i} \tag{3.33}$$

where the quantities $\mu_{i+\frac{1}{2}}, D_{i+\frac{1}{2}}, \hat{n}_{i+\frac{1}{2}}$ are computed as average values of adjacent nodes i and $i-1$. Substituting Eq. (3.33) in Eq. (3.32), the linearized form at the i-th node is obtained:

$$\underbrace{\frac{D_{i+\frac{1}{2}}}{\Delta x_i}}_{G_{d_{i+\frac{1}{2}}}}(\hat{n}_{i+1} - \hat{n}_i) - \underbrace{\frac{D_{i-\frac{1}{2}}}{\Delta x_{i-1}}}_{G_{d_{i-\frac{1}{2}}}}(\hat{n}_i - \hat{n}_{i-1}) + \mu_{i+\frac{1}{2}}\underbrace{\frac{\hat{n}_{i+1} + \hat{n}_i}{2}}_{g_{md_i}}E_{i+\frac{1}{2}} - \mu_{i-\frac{1}{2}}\underbrace{\frac{\hat{n}_i + \hat{n}_{i-1}}{2}}_{g_{md_{i-1}}}E_{i-\frac{1}{2}}$$
$$= \underbrace{\frac{\Delta x_i}{2}}_{G_{c_i}}R_i + \underbrace{\frac{\Delta x_{i-1}}{2}}_{G_{c_{i-1}}}R_i + \underbrace{\frac{\Delta x_i}{2}}_{C_{d_i}}\frac{d\hat{n}_i}{dt} + \underbrace{\frac{\Delta x_{i-1}}{2}}_{C_{d_{i-1}}}\frac{d\hat{n}_i}{dt} \tag{3.34}$$

Comparing Eq. (3.34) with Eq. (3.19), new time-dependent terms are evidenced. Considering the excess carrier concentrations $\hat{n}$ as equivalent voltages, the corresponding equivalent T-circuit shown in Fig. 3.14 is drawn. Four components are present: the transconductance g_{md} to model drift effects due to the electric field, the conductance G_d to model the drop of the minority carrier concentration along the diffusion path, the conductance G_c to model the recombination proportional to the carrier concentration, and the diffusion capacitance C_d to include delay in minority carrier propagation.

In summary, the spatial dependence of drift-diffusion transport is solved through a discretization of the transport equations along a distributed substrate network,

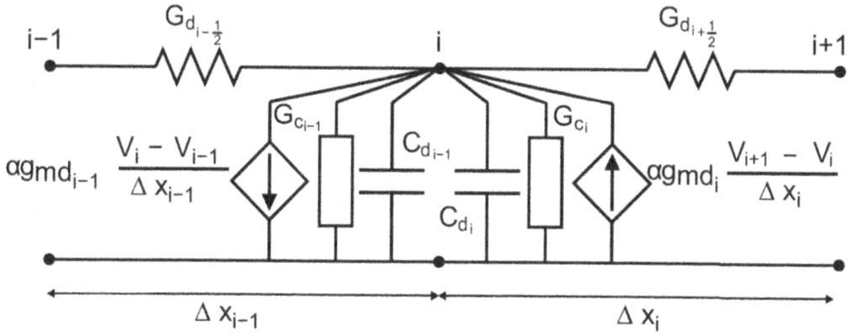

Fig. 3.14 T-equivalent MCC circuit in one dimension ($\alpha = -1$ for holes and $\alpha = +1$ for electrons)

while time-dependent effects are included in the minority carrier circuit MCC through a distributed RC network.

The capacitances C_d do not cover all time-dependent effects in parasitic BJTs because junction capacitive effects are still missing. There are two main contributors to the total AC electrical current: the current from the junction capacitance and the current from the diffusion capacitance. As evidenced in Eq. (3.35) (A is the cross-section area of the junction), such contributions can be expressed as time derivatives of the depletion and diffusion charges, respectively.

$$I_{ac}(V) = \underbrace{A\frac{dQ_p}{dt} - A\frac{dQ_n}{dt}}_{\text{diffusion charges}} + \underbrace{A\frac{dQ_j}{dt}}_{\text{depletion charge}} \tag{3.35}$$

Diffusion charges (first part of the second term of Eq. (3.35)) strongly depend on the minority carrier concentration provided by the diffusion capacitance C_d component of the continuity equation (3.34), while the depletion charge reverts to the classical junction capacitance C_j model for diodes. In the next sections, the modeling of the two contributions are analyzed in detail.

3.4.1.1 Diffusion Capacitance

In forward-biased junctions, minority carriers are injected in the P- and N-doped adjacent layers, and their distributions are sketched in Fig. 3.15. The minority carrier charge densities are space and time-variant, and they contribute as a diffusion

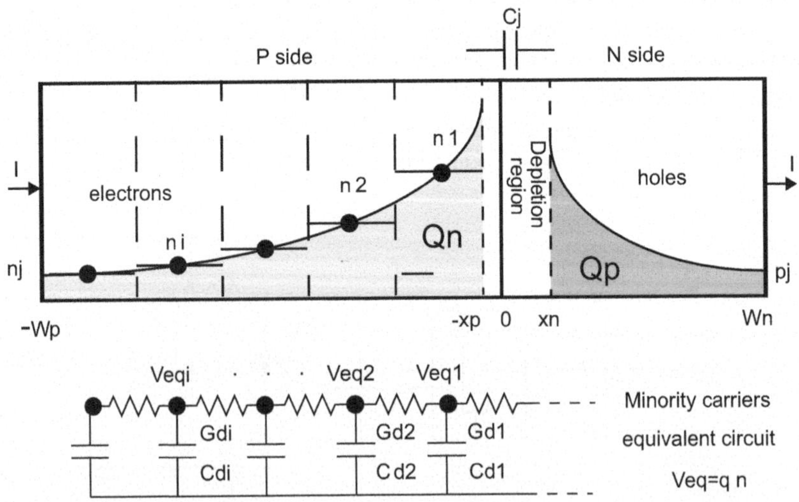

Fig. 3.15 Minority carriers diffusing in a diode with general boundary conditions n_j and p_j and the equivalent approximation with a distributed RC network for the MCC

capacitance in the forward bias condition. Equation (3.36) gives the total injected charge through the integral of the excess charge density in the quasi-neutral regions of the junction (Q_p in the N-side and Q_n in the P-side).

$$Q_p = +q \int_{x_n}^{W_n} \hat{p}(x)\, dx \quad Q_n = -q \int_{-W_p}^{-x_p} \hat{n}(x)\, dx \tag{3.36}$$

Analytical expressions for the carrier distributions $n(x), p(x)$ are available for low applied voltages only, where the excess carrier distributions are essentially driven by diffusion processes. Considering the total recombination of the minority carriers at the physical boundaries ($p_j = 0, n_j = 0$), an exponential voltage-dependence expression for $C_d(V)$ is obtained [21]. However, in high forward bias conditions, drift terms deeply impact the spatial distribution of the minority carriers, and the total diffusion capacitance $C_d(V) = \frac{dQ}{dV}$ cannot be implemented in a closed-form equation. Moreover, for parasitic BJTs, also the voltage dependence of the boundary conditions for the minority carriers p_j, n_j is needed.

To overcome this problems, the integrals in Eq. (3.36) are computed with the rectangle method, which is the same as using an equivalent RC distributed network for the minority carriers. The diffusion capacitance C_d for the MCC uses the minority carrier concentration at node i with the additional terminal voltage $V_{eq,i}$. By connecting together different lumped devices, the circuit simulator computes the equivalent voltages from the boundary conditions p_j, n_j and derives the time-variant terms $C_{d,i} = A\Delta x_i \frac{dV_{eq,i}}{dt}$. Note that this implementation of diffusion capacitances does not require any additional parameters in the model as the time-dependent formulation of the continuity equations use the minority carrier lifetime parameter τ of the model.

The C_d capacitances are included only in the MCC. However, at the junction nodes, the time-dependent of minority charges is coupled to the TCC by the I_{tot} element, introducing a time-dependent current that propagates through the substrate.

3.4.1.2 Junction Capacitance

The second capacitive element is the junction capacitance accounting for varying charges in the space-charge region. The depletion region width x_d is modulated by the applied voltage in a nonlinear way, also depending on the junction doping profile. A general space-dependent doping profile $N(x)$ of the form (3.37) ($x = 0$ corresponds to the junction depth) is considered:

$$N(x) = \begin{cases} D_1 x^{m-2} & \text{for } x < 0 \\ D_2 x^{m-2} & \text{for } x \geq 0 \end{cases} \tag{3.37}$$

D_1, D_2 are constant parameters, while the gradient coefficient m is equal to 2 for the case of an abrupt doping profile and 3 for a linear doping profile. The solution of the

Poisson's equation (considering complete ionization) leads to Eq. (3.38) that links the depletion width to the junction potential drop V_j [22]:

$$x_d(V_j) = \left(\frac{\varepsilon_s(V_{bi} - V_j)m}{q}\right)^{\frac{1}{m}} \left(D_1^{-\frac{1}{m-1}} + D_2^{-\frac{1}{m-1}}\right)^{1-\frac{1}{m}} = x_0(V_{bi} - V_j)^{\frac{1}{m}} \tag{3.38}$$

where ε_s is the silicon permittivity and V_{bi} the built-in potential. Therefore, three model parameters are needed: the grading coefficient m, the built-in potential V_{bi}, and a constant x_0 depending on the doping profile. These parameters are equivalent to the standard diode parameters in the junction capacitance equation $C_j(V_j)$ of spice-diode models since it does not depend on the minority carrier boundary conditions.

$$C_j(V_j) = \frac{C_j}{\left(1 - \frac{V_j}{V_{bi}}\right)^m} \tag{3.39}$$

It is worth to note that Eq. (3.39) suffers from a discontinuity when $V_j \simeq V_{bi}$. In simulators, a linearization is usually done for high voltages. A different circuit implementation is suggested for the substrate model by using a time-varying charge $Q_j(V_j)$ given by Eq. (3.40). This expression does not diverge any more in forward bias conditions ($V_j \geq V_{bi}$) because the depletion width x_d rapidly tends to zero, avoiding numerical issues when implementing C_j.

$$Q_j = \frac{qAx_d^{m-1}}{(m-1)\left(D_1^{-\frac{1}{m-1}} + D_2^{-\frac{1}{m-1}}\right)^{m-1}} = \frac{A\varepsilon_s m}{(1-m)x_0}(V_{bi} - V_j)^{1-\frac{1}{m}} \tag{3.40}$$

The $\frac{dQ_j}{dt}$ is implemented in the diode TCC as an additional nonlinear current source in parallel with the I_{tot} current of Fig. 3.9, covering all junction capacitance-related effects.

3.4.2 Deep Trenches and Epi-layer Model

As described in Sect. 2.2, several technologies are available for HV ICs. The EPFL substrate model can be applied to any technology option based on silicon (or any other semiconductor). The key features of this modeling approach are the meshing concept that takes into account all the geometry variations and the lumped components that use specific technology parameters.

For specific technology options, some modifications of the meshing algorithm have to be considered. Example in Fig. 3.16 reports different cross sections of a parasitic lateral NPN BJT. The structure (a) consists of two n-wells in a low- and

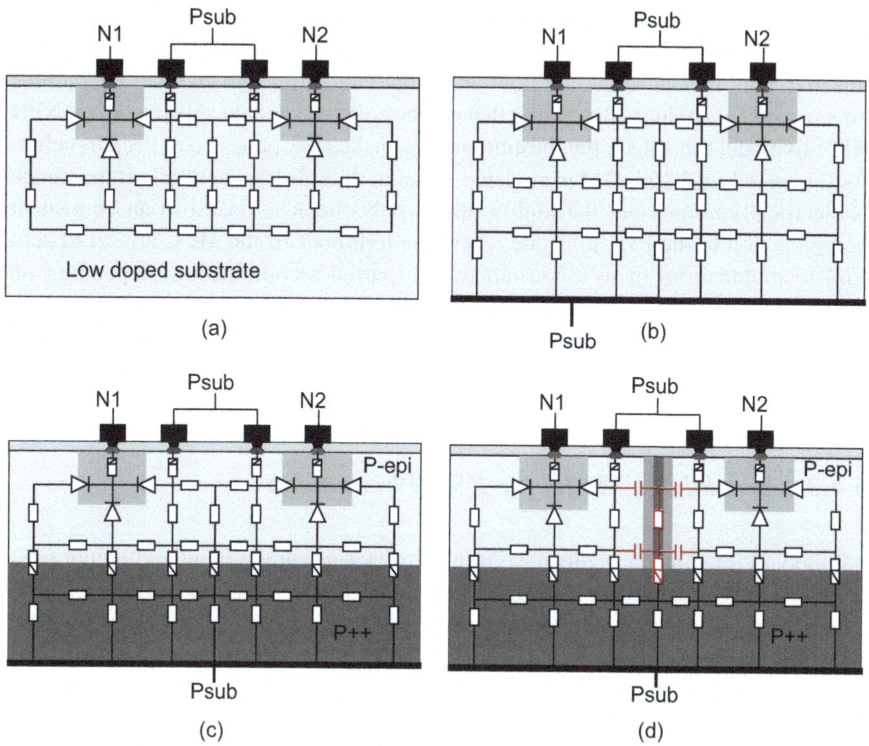

Fig. 3.16 Cross section and equivalent substrate network of two n-wells in different technologies: (**a**) low-doped substrate; (**b**) backside contact; (**c**) epi-layer on high-doped substrate; (**d**) DTI technology with trench modeling

uniformly doped p-type substrate, like in high-voltage CMOS technologies [3]. To have a uniform substrate contact, it is common to use a backside metallization as shown in Fig. 3.16b. In this case, the equivalent substrate network includes an additional row of P resistors to model the backside connection, assuming that there is no Schottky barrier [23].

In other HV processes, a thin lightly P-doped epitaxial layer is grown on a heavily P++-doped substrate; see Fig. 3.16c. In this case, the P–/P++ doping discontinuity completely modifies the minority carrier transport. Due to the barrier effect at the P–/P++ interface [24], minority carriers are induced to flow laterally in the P– region, and the high recombination rate in the P++ substrate reduces the coupling effects due to the lateral NPN BJT. The meshing algorithm must introduce one or more homojunctions at the P–/P++ substrate discontinuity as shown in the equivalent network of Fig. 3.16c. The same concept applies for technologies using N+ buried layers aiming at reducing the gain of parasitic vertical PNP BJT.

Finally, many BCD processes make use of dielectric isolation to further minimize the substrate couplings from minority carriers. For instance, the simplified cross section of a DTI technology option is reported in Fig. 3.16d between two high-voltage n-wells in a P−/P++ substrate. This technology solution forces electrons injected into the substrate to flow into the P++ layer where they recombine effectively. This reduces the charge that can be collected and disable the lateral NPN BJT. To model this effect, the meshing must consider the presence of deep trenches. As shown in Fig. 3.16d, DTI is modeled as cut in the substrate network. If the trench is electrically connected, the highly doped polysilicon is linked to an equivalent homojunction to the P++ substrate where the terminals of the MCC are set to zero (full recombination) or to a constant value (partial recombination) depending on the technology process. Also required are coupling capacitances through the trench oxide. Then, the overall electrical model includes capacitances between adjacent TCC terminals while MCC terminals are left floating.

3.4.3 Breakdown Model for ESD Phenomena

All modern integrated circuits have protection devices at the input and output pads of the chip in order to avoid ESD failures. The ESD protections are designed to clamp voltages and to maintain transient currents at safe levels [25]. These special structures activate only at ESD events absorbing the energy in some nanoseconds to avoid damages. In most cases, the triggering process is avalanche breakdown of reverse-biased junctions (during normal operation).

Typical protection devices are diodes, BJTs, and NMOS transistors with grounded gate (ggNMOS). From a simulation point of view, the breakdown is not included in the spice models of diodes and transistors; thus, ESD protections cannot be simulated without some specific macromodels [26]. Moreover, many ESD devices show a snapback behavior which results in convergence issue for circuit simulators because of the multiple solutions that exist for a given voltage bias [27].

The lack for a simulation model makes it difficult to estimate ESD protection effectiveness, i.e., the ability to protect the circuit implemented on the same IC. Metal or substrate couplings exist between the ESD devices and the other circuits, meaning that ESD currents could spread in the silicon substrate, reducing the robustness of the protection and generating a failure somewhere else in the chip [28]. The substrate model is the solution to address these issues, but it should be extended to include important features involved during ESD events such as short timings, high-current densities, and high-voltage peaks. Three main aspects are usually required for ESD modeling:

- **High currents:** when high-current density is flowing in a silicon device, local electric fields are important to determine drift currents that dominate diffusion currents. Under high injection, when the injected minority carrier concentration at junctions is comparable with the doping concentration, the local resistivity of the silicon is modulated.

- **Thermal effects:** with high energy flowing inside the substrate, temperature gradients are generated, and a self-heating model should be included to allow non-isothermal simulations. Electrothermal simulations are used to estimate the critical current density that may degrade the semiconductor.
- **Junction breakdown:** for high reverse bias voltages, the electric field in the depletion region reaches critical values that trigger avalanche breakdown phenomena. These mechanisms are used to activate local electrical feedbacks in ESD protection devices.

Thermal effects can be neglected in first analysis considering only isothermal conditions. Temperature dependencies of silicon parameters are usually available only for ranges below 600 K. Also TCAD simulations are not reliable to exploit the thermal failures near the melting point of silicon, i.e., 1680 K [29]. Due to this limitation, no electrothermal models will be discussed.

The EPFL substrate model is a good candidate to simulate ESD-substrate parasitic couplings which can unintentionally trigger a latch-up. The model naturally takes into account the distributed geometrical effects that are included in the macro-modeling of ESD protection devices [30]. Also high-injection effects and substrate conductivity modulation are already included in the GLD. Nevertheless, a model for the avalanche breakdown is still required. The model being based on PN junctions, once the physics of breakdown mechanisms is included in the EPFL diode, bipolar transistors and more complex structures such as thyristors can also be simulated during ESD events.

Extending the model for breakdown asks for in-depth analysis of the reverse bias condition of PN junctions. In reverse bias, the applied voltage V_j drops entirely across the depletion region. A critical voltage exists where the electric field is so high that it can induce avalanche generation of electron-hole pairs, resulting in injection of majority carriers (electrons are injected in the N-side and holes in the P-side) and in a steep increase of the current density.

The value of the critical electric field responsible for avalanche breakdown is highly dependent on the junction geometry (rounded shapes avoid electric field crowding, for instance) and on the doping profiles (low-doped regions lead to higher breakdown voltages). These are usually obtained from measurements on working devices. The avalanche current passing through the junction can be modeled using a current multiplication factor M that depends on the impact ionization integral [21]. No closed-form solution for this integral exists, and the empirical Miller's formula is used [31] instead

$$M = \frac{1}{1 - \left(\frac{-V_j}{BV}\right)^{\alpha}} \tag{3.41}$$

where V_j is the applied voltage across the junction, BV is the breakdown voltage, and $\alpha = 4\ldots6$ is a technology-dependent coefficient. This expression is useful to estimate the avalanche breakdown currents; however, it suffers from a mathematical discontinuity around BV. Then, it cannot be implemented in circuit simulators because of converge problems. A continuous function rapidly increasing at BV is preferred [32]:

$$M = \exp\left(\left(\frac{-V_j}{BV}\right)^{\alpha}\right) \tag{3.42}$$

This expression is valid in reverse bias ($V_j < 0$) and introduces two parameters BV and $\alpha > 1$ that must be extracted for a given technological process. This formula is quite general and can be used to include also other mechanisms such as tunneling.

To include breakdown features in the model requires to multiply the injected current I_{tot} in the TCC of the EPFL diode with the multiplication factor M. Since I_{tot} is coupled to the injected concentration of minority carriers (cf. $V_{eq,n}$, $V_{eq,p}$), when the avalanche phenomena is triggered, the junction current increases through the injection of majority carriers. At the same time, the minority carrier concentration is also modified.

3.5 Conclusions

The parasitic substrate current can be modeled by discretization of the drift-diffusion equations using FDM. This approach translates the mathematical expressions in terms of two equivalent circuits: the TCC and the MCC. The first one propagates real currents and voltages, while in the latter equivalent currents and voltages related to minority carriers are defined. Combining the two coupled sub-circuits, lumped components are defined: the EPFL resistor, the EPFL diode for PN junction interfaces, and the EPFL homojunction for modeling contacts effects at PP+ and NN+ interfaces. These three elements constitute the generalized lumped devices (GLD) that, following a proper interconnected scheme, can simulate substrate currents due to parasitic BJT, for instance. In this modeling methodology, a meshing concept is required to divide the space in cuboids where the different lumped components are instantiated. The meshing algorithm can be adapted to various HV technological options, which make the modeling concept very flexible. Compared to TCAD-based substrate modeling solutions, simulation of the substrate parasitic network composed of GLD is fully spice-compatible and relatively fast and requires a minimum amount of computational resources.

References

1. A.F. Franz, G.A. Franz, S. Selberherr, C. Ringhofer, P. Markowich, Finite boxes - a generalization of the finite-difference method suitable for semiconductor device simulation. IEEE Trans. Electron Devices **30**(9), 1070–1082 (1983)

2. V. Binet, Y. Savaria, M. Meunier, Y. Gagnon, Modeling the substrate noise injected by a DC-DC converter, in *IEEE International Symposium on Circuits and Systems (ISCAS)* (May 2007), pp. 309–312
3. M. Schrems, M. Knaipp, H. Enichlmair, V. Vescoli, R. Minixhofer, E. Seebacher, F. Leisenberger, E. Wachmann, G. Schatzberger, H. Gensinger, Scalable high voltage CMOS technology for smart power and sensor applications. e & i Elektrotechnik und Informationstechnik **125**(4), 109–117 (2008)
4. F. Lo Conte, J.-M. Sallese, M. Pastre, F. Krummenacher, M. Kayal, Global modeling strategy of parasitic coupled currents induced by minority-carrier propagation in semiconductor substrates. IEEE Trans. Electron Devices **57**(1), 263–272 (2010)
5. F. Lo Conte, J.-M. Sallese, M. Kayal, Modeling methodology of high-voltage substrate minority and majority carrier injections, in *Proceedings of the European Solid-State Device Research Conference (ESSDERC)* (Sept 2010), pp. 194–197
6. F.J.R. Clement, E. Zysman, M. Kayal, M. Declercq, LAYIN: toward a global solution for parasitic coupling modeling and visualization, in *Proceedings of the IEEE Custom Integrated Circuits Conference (CICC)* (May 1994), pp. 537–540
7. Open Verilog International, *Verilog-AMS Language Reference Manual: Analog & Mixed-Signal Extension to Verilog HDL; Version 2.0* (Open Verilog International, 2000). https://books.google.ch/books?id=iXWvPgAACAAJ
8. C. Christopoulos, *The Transmission-Line Modeling Method TLM* (IEEE Press, New York, 1995)
9. M.Y. Al-Zeben, A.H.M. Saleh, M.A. Al-Omar, TLM modelling of diffusion, drift and recombination of charge carriers in semiconductors. Int. J. Numer. Modell. Electron. Networks Devices Fields **5**(4), 219–225 (1992)
10. N.D. Jankovic, T.V. Pesic, J.P. Karamarkovic, 1D physical non-quasi static BJT circuit model based on the equivalent transmission line analysis. J. Comput. Electron. **3**(1), 13–24 (2004)
11. S. Selberherr, *Analysis and Simulation of Semiconductor Devices* (Springer Science & Business Media, 2012)
12. B.R. Stanisic, N.K. Verghese, R.A. Rutenbar, L.R. Carley, D.J. Allstot, Addressing substrate coupling in mixed-mode ICs: simulation and power distribution synthesis. IEEE J. Solid State Circuits **29**(3), 226–238 (1994)
13. P. Buccella, C. Stefanucci, J.-M. Sallese, M. Kayal, Spice simulation of substrate minority carriers propagation with equivalent electrical circuit, in *Proceedings of the 21st International Conference Mixed Design of Integrated Circuits Systems (MIXDES)* (June 2014), pp. 347–350
14. M.P. Beddoes, Some comments on Linvill's lumped models for semiconductor devices. Proc. Inst. Electr. Eng. **112**(8), 1509–1514 (1965)
15. F. Lo Conte, J.-M. Sallese, M. Kayal, Smart power IC simulation of substrate coupled current due to majority and minority carriers transports, in *IEEE International Conference on IC Design and Technology (ICICDT)* (June 2010), pp. 168–171
16. T. Misawa, A note on the extended theory of the junction transistor. J. Phys. Soc. Jpn. **11**(7), 728–739 (1956)
17. C.-T. Sah, R.N. Noyce, W. Shockley, Carrier generation and recombination in P-N junctions and P-N junction characteristics. Proc. IRE **45**(9), 1228–1243 (1957)
18. S. Li, Y. Fu, Smart power technology and power semiconductor devices, in *3D TCAD Simulation for Semiconductor Processes, Devices and Optoelectronics* (Springer, New York, 2012), pp. 187–236
19. J. Oehmen, L. Hedrich, M. Olbrich, E. Barke, A methodology for modeling lateral parasitic transistors in smart power ICs, in *IEEE Behavioral Modeling and Simulation (BMAS)* (Sept 2005), pp. 19–24
20. C. Stefanucci, P. Buccella, M. Kayal, J.M. Sallese, Impact of enhanced contact doping on minority carriers diffusion currents, in *10th Conference on Ph. D. Research in Microelectronics and Electronics (PRIME), 2014* (IEEE, New York, 2014), pp. 1–4
21. S.M. Sze, K.K. Ng, *Physics of Semiconductor Devices* (Wiley-Interscience, London, 2006)

22. V. Milovanovic, H. Zimmermann, Analytical analysis of a p-n junction with arbitrary shaped doping profile, in *28th International Conference on Microelectronics (MIEL)* (May 2012), pp. 73–76
23. M. Schenkel, Substrate current effects in smart power ICs. PhD thesis, ETH Zürich, Nr. 14925, 2003
24. R. Stella, S. Favilla, G. Croce, Novel achievements in the understanding and suppression of parasitic minority carrier currents in P- epitaxy/P++ substrate smart power technologies, in *Proceedings of the IEEE 16th International Symposium on Power Semiconductor Devices and ICs (ISPSD)* (May 2004), pp. 423–426
25. A.Z.H. Wang, *On-Chip ESD Protection for Integrated Circuits, An IC Design Perspective* (Kluwer Academic Publishers, Dordrecht, 2006)
26. J. Li, S. Joshi, R. Barnes, E. Rosenbaum, Compact modeling of on-chip ESD protection devices using Verilog-A. IEEE Trans. Comput. Aided Des. Integr. Circuits Syst. **25**(6), 1047–1063 (2006)
27. L. Wei, C.E. Gill, W. Li, R. Wang, M. Zunino, A convergence robust method to model snapback for ESD simulation, in *International Semiconductor Conference (CAS)*, vol. 2 (Oct 2011), pp. 369–372
28. A. Gendron, C. Gill, C. Aykroyd, C. Zhan, Techniques to prevent substrate injection induced failure during ESD events in automotive applications, in *Proceedings of the IEEE 23rd International Symposium on Power Semiconductor Devices and ICs (ISPSD)* (May 2011), pp. 192–195
29. C. Salamero, N. Nolhier, A. Gendron, M. Bafleur, P. Besse, M. Zecri, TCAD methodology for ESD robustness prediction of smart power ESD devices. IEEE Trans. Device Mater. Reliab. **6**(3), 399–407 (2006)
30. K.-H. Meng, E. Rosenbaum, Layout-aware, distributed, compact model for multi-finger MOSFETs operating under ESD conditions, in *Electrical Overstress/Electrostatic Discharge Symposium (EOS/ESD)* (Sept 2013), pp. 1–8
31. S.L. Miller, Ionization rates for holes and electrons in silicon. Phys. Rev. **105**, 1246–1249 (1957)
32. S.L. Lim, X.Y. Zhang, Z. Yu, S. Beebe, R.W. Dutton, A computationally stable quasi-empirical compact model for the simulation of MOS breakdown in ESD-protection circuit design, in *International Conference on Simulation of Semiconductor Processes and Devices (SISPAD)* (Sept 1997), pp. 161–164

Chapter 4
TCAD Validation of the Model

4.1 High Injection Effects in Diodes

Figure 4.1 shows the simplest diode structure with given geometrical and doping parameters with its equivalent "enhanced" model. The homojunctions at the boundaries are required to account for the N+ and P+ contacts. A low-doped p-substrate has been considered as a case study in order to include the resistivity of the diode.

For a cross-section area of $2.5\,\mu m^2$, the expected substrate resistance estimated from the doping concentration of the structure is about $780\,k\Omega$. At high forward bias, the current should be limited by this large resistivity. However, simulations reveal that several hundreds of milliamperes can flow when the applied voltage is greater than 1 V.

Figure 4.2 shows the current-voltage characteristics of the diode with a forward bias of up to 2 V for TCAD and VerilogA simulations at different temperatures. Both for TCAD and GLD models, Arora's mobility relations and Sharfetter's relations for lifetime are implemented to include doping and temperature dependencies (see Appendix A). In the VerilogA simulations, ohmic contacts are assumed to have zero resistance in order to check the intrinsic conductivity modulation effect. However, a contact resistance due to the metallization could limit the current density in real devices.

Conductivity modulation takes place at high-injection regimes. If the coupling mechanisms in Fig. 3.4 are disabled, the classical doping-dependent resistivity is simulated. The standard doping-dependent conductivity model overestimates the diode resistance and results in orders of magnitude mismatch in the current densities.

The effect of conductivity modulation on the diode series resistance is reported in Fig. 4.3. The p-type substrate is affected first by the high carrier injection. In low injection several $k\Omega$ are simulated as expected from the low-doping concentrations N_a or N_d, whereas under high voltages (high injection), the resistance drops down to

P. Buccella et al., *Parasitic Substrate Coupling in High Voltage Integrated Circuits*, Analog Circuits and Signal Processing, https://doi.org/10.1007/978-3-319-74382-0_4

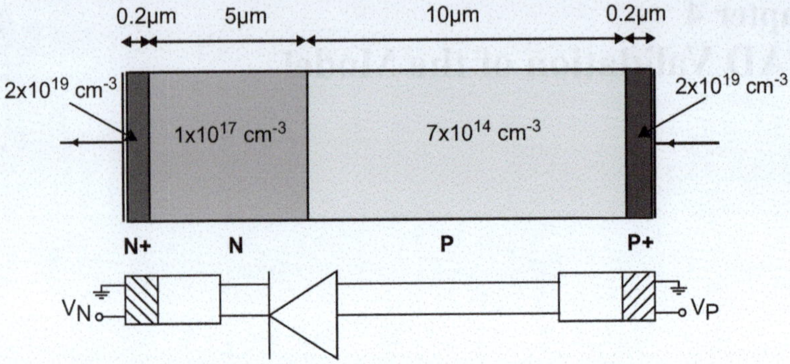

Fig. 4.1 Diode with N+ and P+ contacts (top) and the simulated equivalent circuit (bottom)

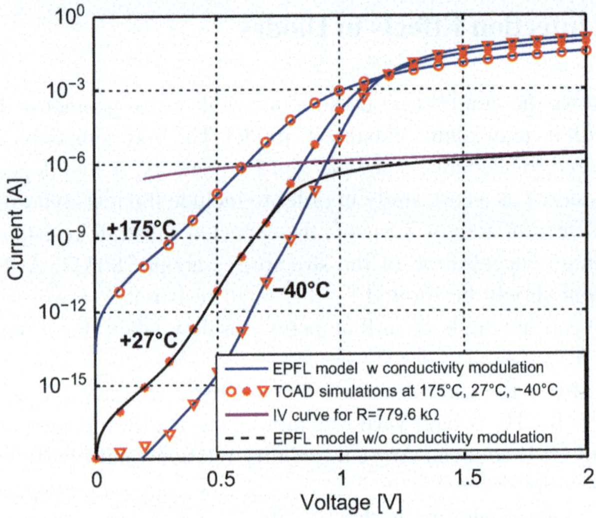

Fig. 4.2 Simulated IV characteristic of a diode at different temperatures

a few ohms only. The same conductivity modulation is present in BJTs [13] where the base resistance is degraded.

But the conductivity is also modulated by the injection of minority carriers as shown in Fig. 4.3 where ten meshing nodes are considered for the p-type substrate (the applied voltage is 1.2 V). Close to the PN junction, the resistance is lowered, while next to the PP+ junction, it is increased. This means that electrons are more concentrated at the PP+ homojunction than at the PN junction. This counterintuitive effect comes from the PP+ interface where electrons accumulate. When the P+ is removed, the opposite behavior is predicted, and the injected electrons accumulate at the PN junction. These simulations prove the importance of homojunctions in modeling minority carriers.

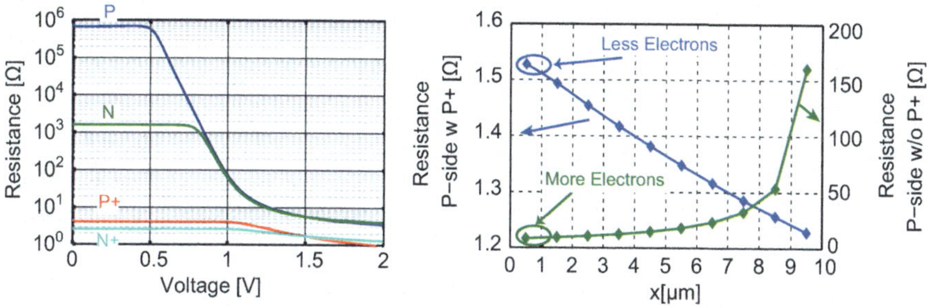

Fig. 4.3 (Left) Simulated resistance variation due to the injection of minority carriers at different bias levels. (Right) P-substrate conductivity modulation at 1.2 V with and without P+ contact

Moreover, if more electrons are present at PP+ interface under high-injection conditions, the gradient of minority carriers is reversed, and the diffusion current becomes negative, as confirmed by simulations. However, in this case the drift component is still the dominant contribution. In addition, the simulated electron current at 1.5 V is 19 mA, while the hole current is only 8 mA, mainly because the mobility of electrons is higher than the hole's mobility.

To validate the capacitance model in the diode element, the substrate length is changed to keep 500 nm N+ and P+ contacts at the extremities. Two resulting diodes are considered: a short diode with $W_p = 10\,\mu\text{m}$ emulating the distance between a well and a P+ contact and a long diode with $W_p = 500\,\mu\text{m}$ emulating a wafer backside contact beneath a well. In this example, the substrate doping is $N_a = 10^{16}\,\text{cm}^{-3}$.

Figure 4.4 shows the simulated CV characteristics for the two devices. In reverse bias the model simulates the junction capacitance as in standard SPICE models. In forward bias, the capacitive effects related to the drift-diffusion of the minority carriers are evidenced by the exponential increase of the diffusion capacitance. Note that after reaching a maximum, the capacitance decreases, then becomes negative, thus behaving as an inductance [11].

The diode's inductive behavior is a consequence of the high injection of minority carriers and the related modulation of the conductivity [4, 5]. Classical diode models simulate this effect by introducing an extrinsic lumped inductance to the diode compact model. Here, the EPFL lumped device predicts such an inductive behavior as it accounts for transport delay for minority carriers. Unlike the junction capacitance, the diffusion capacitance C_d is frequency-dependent, and the impedance will depend on the diode length. The distributed nature of the capacitive network can simulate frequency domain simulations up to 100 MHz.

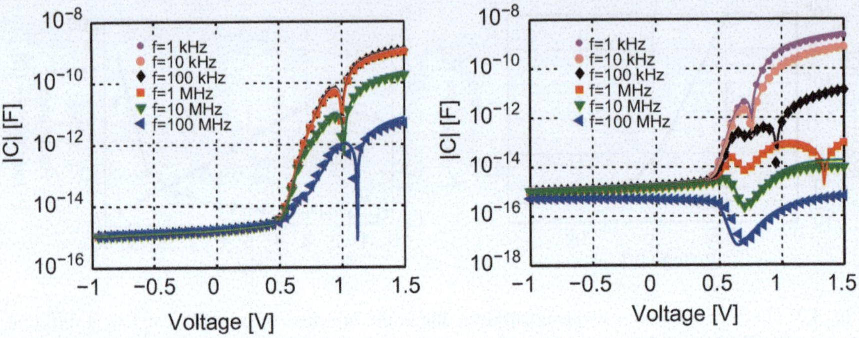

Fig. 4.4 CV plot from AC simulations at different frequencies for a short diode (left) and a long diode (right)

4.2 NPN Lateral Bipolar Junction Transistor

Bipolar transistors can be simulated by interconnecting different lumped devices. We consider a typical aggressor-victim configuration in HV ICs between two similar n-wells with average donor doping concentration $N_d = 10^{17}\,\text{cm}^{-3}$. One well, the emitter N_1, is biased belowground and injects electrons in the p-substrate, while the other N well (the collector N_2) is reverse biased at 50 V (supply voltage) and collects the parasitic current. The low-doped p-substrate (equivalent doping of $N_a = 4\times10^{15}\,\text{cm}^{-3}$) is set at the reference potential 0 V and acts as the base of the parasitic lateral BJT. This configuration can be simulated with an equivalent NPN BJT where the parameters and the equivalent GLD circuit model are reported in Fig. 4.5. For completeness P+ and N+ contacts doped at $10^{19}\,\text{cm}^{-3}$ are also included.

Simulations of emitter, collector, and base currents are reported in the Gummel plot of Fig. 4.6 when the two wells are separated by a $d = 20\,\mu\text{m}$ and when the base contact is at the center. Simulation from TCAD and from the EPFL substrate model is in good agreement from low- to high-current levels. Around 1.2 V the base current becomes larger than the collector current because drift mechanisms dominate at these relatively high voltages. The induced gain degradation is evidenced in the high-injection β roll-off. If the P+ contact is removed and 0 V is directly applied to the low-doped substrate, the simulation reveals a large mismatch for the gain β. The presence of the PP+ homojunction does not allow electrons to fully recombine, and they pass through the base toward the collector. This effect is called Minority carrier mirror (MCM). Again, modeling homojunctions is fundamental when modeling minority carrier propagation.

When the distance between the two wells is doubled ($d = 40\,\mu\text{m}$), the bipolar effect is weakened. In this case the meshing algorithm instantiates more diffusion resistors in the base equivalent circuit. The simulations of the new parasitic NPN BJT (the base contact is still in the center between the n-wells) show a 30%

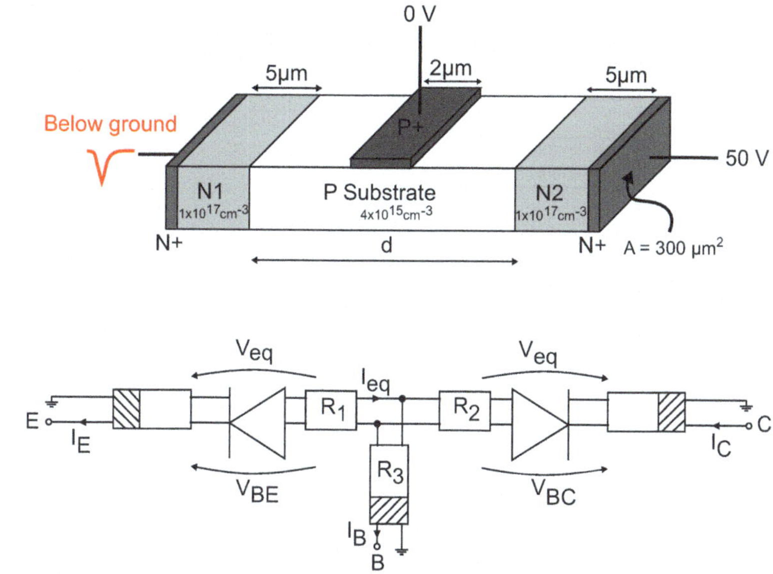

Fig. 4.5 Simulated parasitic NPN BJT (top) with equivalent circuit (bottom)

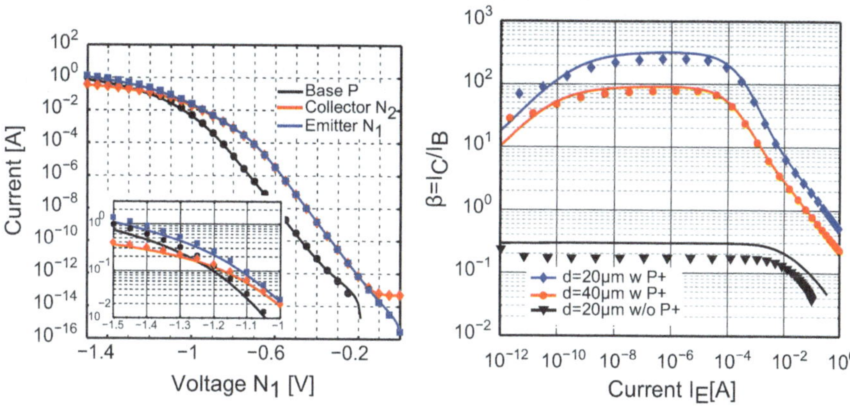

Fig. 4.6 (Left) NPN BJT Gummel plot simulation for a base length $d = 20\,\mu\text{m}$. (Right) Simulated gain β for different base lengths

reduction in the gain. Figure 4.7 shows the simulated base transport factor α for the two devices, a parameter commonly used to characterize parasitic substrate couplings. In the low-injection regime ($I_E < 1\,\text{mA}$), the distance has no impact, and the two wells remain strongly coupled ($\alpha \simeq 1$). However, in high-injection ($I_E > 1\,\text{mA}$), the increase in the base resistance reduces the coupling. The base resistance composed by interconnecting R_1, R_2, and R_3 (see equivalent circuit in

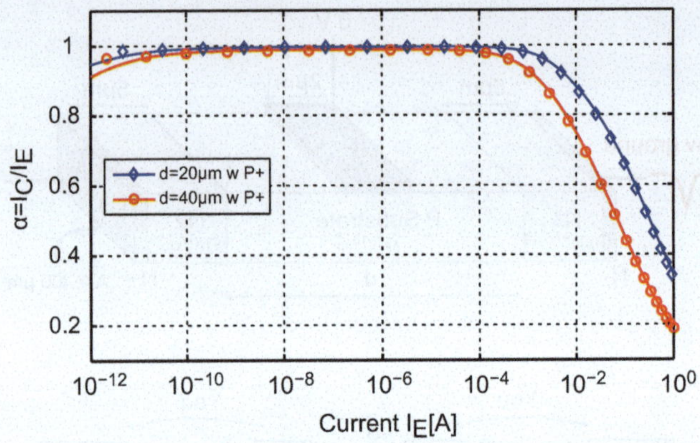

Fig. 4.7 Simulated NPN BJT base transport factor α for two different base lengths

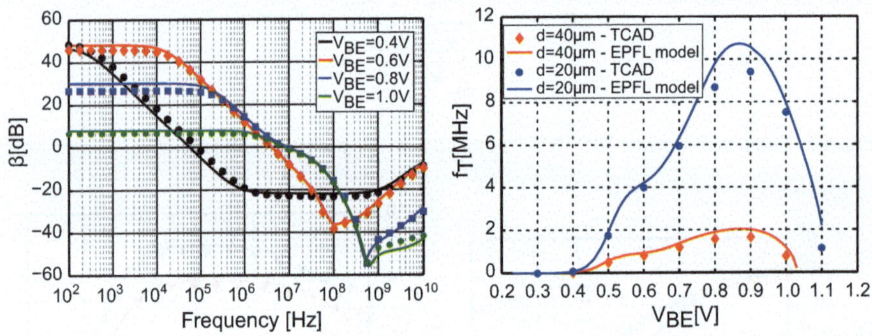

Fig. 4.8 (Left) Frequency variation of current gain for the BJT of Fig. 4.5. (Right) Cutoff frequency simulation of parasitic NPN BJT with different base lengths

Fig. 4.5) is well modeled. Compared to TCAD, an improvement in simulation time of about 1000 is reported with the SPICE-based modeling, meaning that it is possible to use the tool for fast optimization of a lateral bipolar transistor as well.

Furthermore, since all the capacitive components are embedded in the EPFL diode, no extra base-emitter C_{BE}, base-collector C_{BC}, or collector-emitter C_{CE} capacitances are needed to take into account frequency-dependent characteristics of a BJT. AC simulations have been performed by applying a small signal to the base-emitter junction of the device with $d = 20\,\mu\mathrm{m}$, keeping the collector reverse biased at 50 V. The frequency dependence of β is reported in Fig. 4.8 for different forward bias voltages. The results of the equivalent GLD circuit match with TCAD, and the simulated current gain reaches 0 dB at the cutoff frequency f_T. The variation of f_T is also reported by running simulations for different DC bias. As expected, the cutoff frequency has a maximum and then decreases in high-current regimes due to the β roll-off. The extraction of f_T requires several AC simulations over a wide frequency range. The time required to simulate ten f_T points in TCAD is around 4 h, while only 9 min are required in circuit simulator while processing a greater number of points.

The EPFL lumped devices are able to predict the variation of f_T with the base length as well. Considering the device with $d = 40\,\mu\text{m}$, the delays related to the transport of the minority carriers across the base increases, implying a reduction in f_T. As a general comment, the various parasitic BJTs present in a given IC have not only different gains but also different time delays that can be effectively handled by the EPFL circuit model.

4.3 PNP Vertical Bipolar Junction Transistor

Figure 4.9 shows the cross section of typical Lateral diffused MOS devices (LDMOS) available in HV CMOS technology. Both P- and N-Metal-oxide semiconductor (MOS) have an n-well that isolates the transistors from the substrate. In both devices this well introduces a parasitic vertical PNP BJT. In the case of a PMOS, the parasitic emitter corresponds to the drain of the MOS transistor that is usually connected to the load. Then, the vertical BJT can be activated by injecting a hole current (majority carriers) into the substrate that can generate deleterious potential shifts and failures [8].

A compact model for this device is usually added as part of the compact model of the LDMOS to simulate the parasitic currents for the vertical PNP BJT. If the parasitic BJT model scales with the transistor with W and length L [9], the simulations can predict accurately the injected substrate current but not the local potential shift of the substrate beneath the HVMOS n-wells. For that purpose the substrate resistance must be estimated accurately.

By using the EPFL substrate model, the vertical PNP BJT is "naturally" taken into account during the meshing process and impacts the substrate potential distribution in three dimensions, without the need to introduce a parasitic BJT in the compact model of the LDMOS.

In Fig. 4.10 only the relevant mask layers of the PMOS are used to model the parasitic vertical PNP BJT. The instantiated diodes are of two types: the DN/p-substrate and the DP/DN diode. All the geometrical effects are included in the substrate network obtained from the meshing algorithm since every lumped component has a dedicated area and length. This distributed approach allows to separate the lateral and vertical contributions and is highly scalable with respect

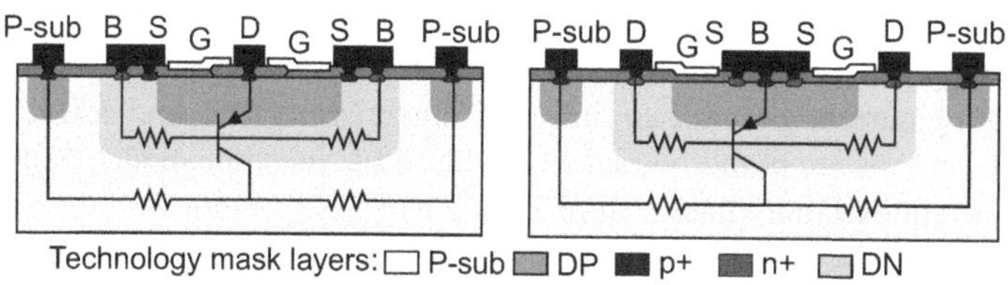

Fig. 4.9 Cross-section view of LDMOS devices with parasitic vertical BJT: PMOS (left) and NMOS (right)

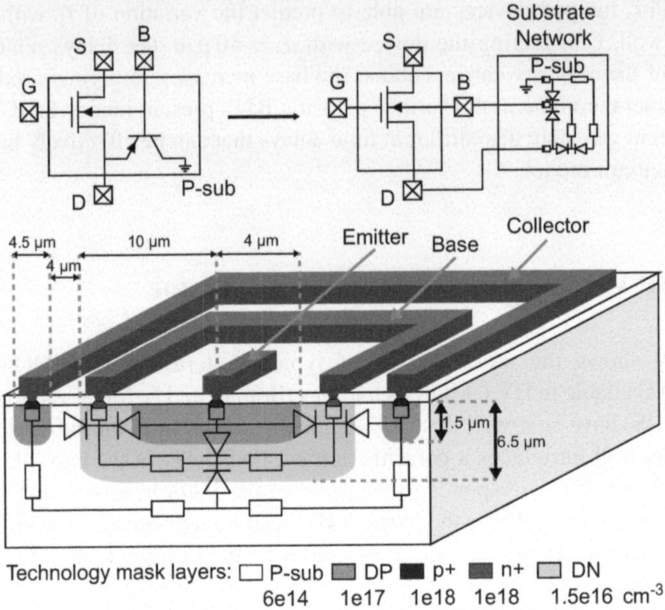

Fig. 4.10 Modeling approach where the parasitic components are separated from the MOS compact model (top). Parasitic vertical PNP BJT with GLD equivalent circuit (bottom)

to the original transistor geometry. The extracted three-dimensional network is equivalent to the Gummel-Poon compact model and can be linked to the compact model of a LDMOS without parasitic elements.

There are two main advantages in using the EPFL substrate modeling approach. The first one is the spatial substrate potential distribution that can be monitored in three dimensions. The second advantage deals with layout optimization. If lateral NPN BJTs are present in the chip layout, these are extracted by the tool, and harmful latch-up configurations can be predicted at early design stages.

Figure 4.11 shows the results obtained from TCAD and from the circuit simulator. A low-doped p-substrate is considered and a box of $700 \times 700 \times 700\,\mu m^3$ is simulated. The depth of the structure is considered to be around 50 µm, emulating a LDMOS with two fingers, and the total width is 100 µm. The substrate (parasitic BJT collector) is set to 0 V with P+ contacts, the n-well (base of PNP BJT) is biased at 12 V, and p-well (emitter of PNP BJT) is swept from 12 V to 12.8 V, emulating an above supply condition of the drain of the PMOS. The Gummel plots with the corresponding current gain at two different temperatures (300 and 400 K) confirm that thermal effects are properly tracked by the EPFL substrate model. Note that 3D simulations required more than 10 h in TCAD while only 1 min with the SPICE 3D netlist.

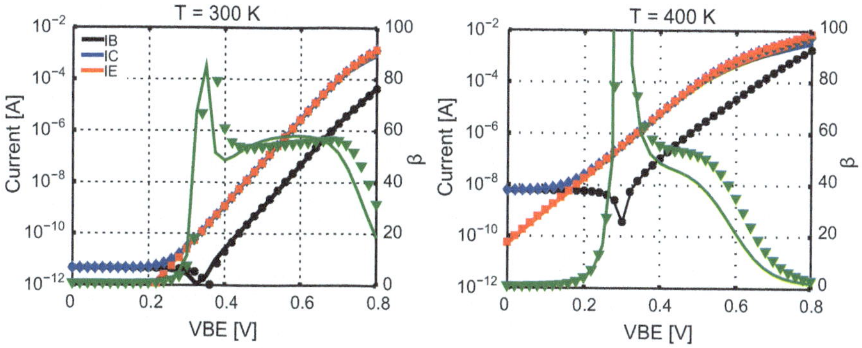

Fig. 4.11 Gummel plot at different temperatures of the vertical PNP BJT of Fig. 4.10. Points correspond to TCAD simulations and continuous lines to GLD model

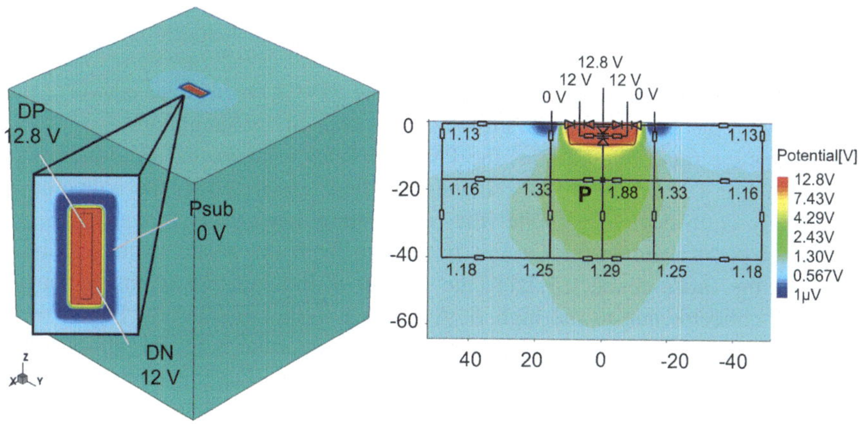

Fig. 4.12 (Left) Three-dimensional potential distribution obtained from TCAD for an injected current of $I_E = 1.4\,\text{mA}$. (Right) Comparison between the potential shift simulated with TCAD (color plot) and with the circuit (numbers correspond to simulated node voltages)

4.3.1 Substrate Potential Shift Simulations

The activation of a vertical PNP BJT injects holes into the substrate that leads to a potential shift. In Fig. 4.12 is reported the simulated substrate potential distribution in TCAD when $V_{\text{BE}} = 800\,\text{mV}$ and $I_E = 1.4\,\text{mA}$. Without a backside contact, the p-type collector ring around the BJT sets the ground voltage (0 V) only at the surface. When the PNP transistor is highly forward biased, the whole substrate potential shifts upward to almost 1 V. This configuration can be detrimental if some sensitive circuit is placed around this device.

The potential distribution is even worse below the n-well. At 18 µm depth, the potential at P in Fig. 4.12 rises to almost 2 V. The high-current gain of the parasitic BJT transistor ($\beta \simeq 55$) leads to this significant substrate potential shift. With this

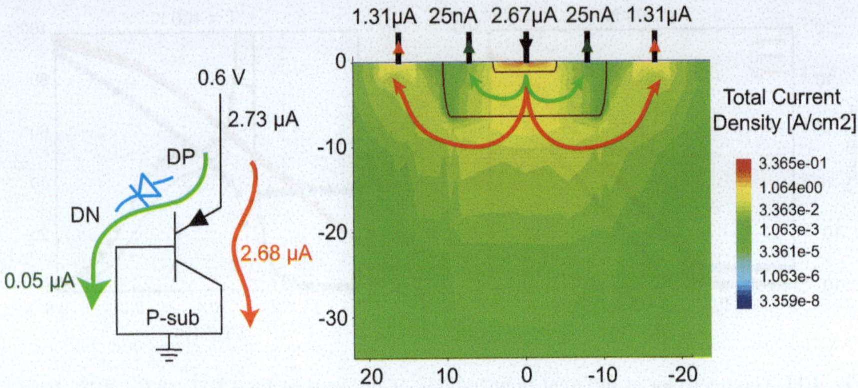

Fig. 4.13 Currents in diode-connected PNP BJT. SPICE simulation (left) and TCAD color plot of the total current density (right)

approach, by changing the node voltages of the 3D GLD substrate network, the de-biasing of the substrate can be investigated within circuit simulators.

In some circuit configurations, the PNP transistor is connected as a diode. As shown in Fig. 4.13, the PNP BJT having the base connected to the collector is electrically equivalent to a diode (the base-emitter junction is forward biased). Nevertheless, in a diode the current is flowing from the anode to the cathode, while in the diode-connected BJT, the current is flowing mostly inside the p-substrate collector due to the high β. According to the simulations, for 600 mV the substrate de-biasing is only 1 mV, but when the current is increased, its value can be much higher and can generate improper operation of nearby circuits.

4.3.2 *Effect of Guard Rings*

The substrate potential can be kept under control with additional p-type contacts and a proper grounding scheme. The simulations reported in Fig. 4.12 clearly show that the 4.5 μm wide p-type guard ring around the device is not sufficient to maintain at 0 V the substrate potential when the parasitic PNP BJT is activated. Once the substrate potential shift is simulated, the aim for the designer is to modify the layout adding additional guard rings to reduce the substrate de-biasing. For example, if the p-type contact ring is extended from 4.5 μm to 14.5 μm, the substrate potential is expected to be lower and better controlled. TCAD simulations (see Fig. 4.14) confirm that instead of almost 1 V of substrate de-biasing obtained when V_E = 12.8 V, the layout modification limits this potential shift to 0.3 V. In addition, this does not affect currents in the vertical PNP BJT, and the injected substrate current remains the same and equal to 1.4 mA.

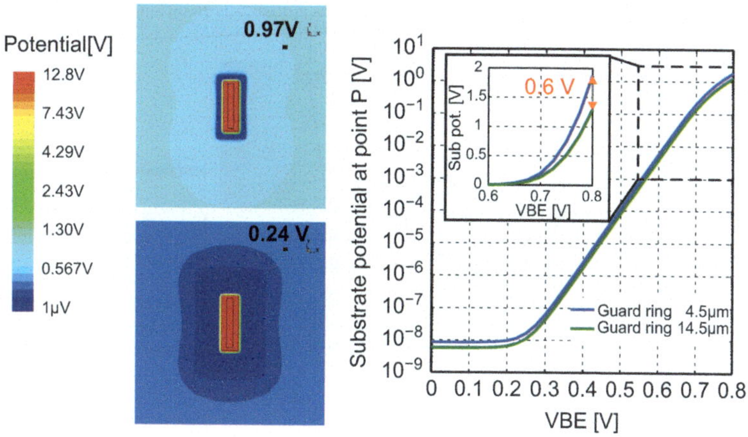

Fig. 4.14 (Left) TCAD surface potential distribution and substrate de-biasing with different guard ring geometries (left 4.5 µm wide, right 14.5 µm wide) for $I_E = 1.4$ mA. (Right) Substrate potential shift beneath the DN well (point P of Fig. 4.12) predicted by the GLD model as a function of biasing and guard ring

Improvement using the second layout configuration is confirmed by the EPFL substrate model combined with the GLD parasitic network. In Fig. 4.14 the substrate potential below the well are shown for the VerilogA model. The difference between the two configurations leads to 600 mV reduction of the substrate de-biasing, as expected from TCAD as well. This demonstrates that the EPFL substrate model can be used to investigate proper grounding scheme around critical parasitic PNP BJT.

4.4 Multi-Collector Transient Couplings

The EPFL substrate model can be used to explore transient substrate couplings in HV ICs. The 2D test structure shown in Fig. 4.15 is a typical aggressor-victim configuration. When the high-voltage MOS drain (e.g., the output stage of an H-Bridge) goes belowground, the corresponding forward-biased n-well injects minority carriers into the substrate and generates a parasitic current. Other lower-voltage-biased wells (e.g., biased at 5 V) act as current collectors, even when placed distant from the aggressor, 70 µm in this case. The test structure is a multi-collector parasitic lateral BJT, where electrons drift and diffuse in the low-doped substrate that is a kind of distributed base set at 0 V.

Transient TCAD simulations are run adopting Gaussian doping profiles for the n-wells ($N_{peak} = 2 \times 10^{17}\,cm^{-3}$) and for the N+ and P+ contacts ($N_{peak} = 5 \times 10^{19}\,cm^{-3}$). Since the GLD is based on average doping concentrations, a calibration of the input parameters is done for the EPFL substrate model.

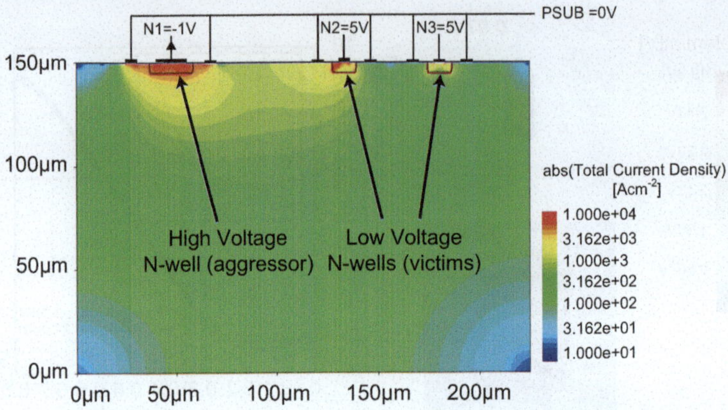

Fig. 4.15 Aggressor-victim configuration between the drain of a high-voltage transistor and low-voltage n-wells inside a low-doped p-substrate (simulated TCAD total current density color plot for $V_{N_1} = -1\text{V}$ belowground condition)

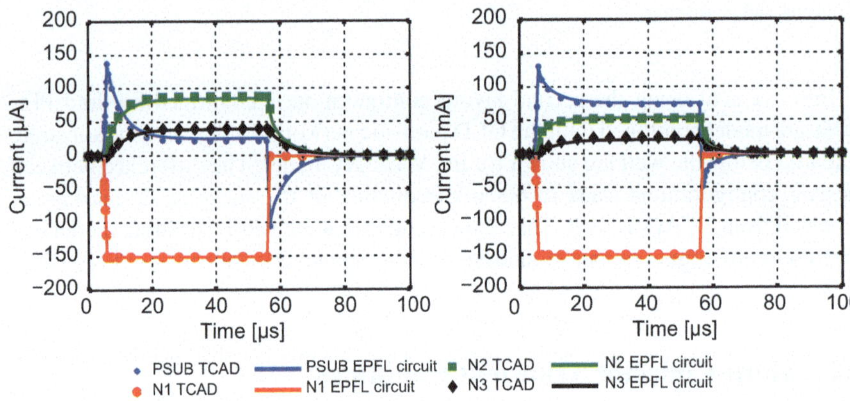

Fig. 4.16 Transient belowground simulation of coupling mechanisms between different wells (cf. Fig. 4.15) for low-injection (left, $V_{N_1} = -0.67\,\text{V}$) and high-injection (right, $V_{N_1} = -1\,\text{V}$) of electrons

A transient stimulus is applied to the HV well N_1 emulating the effect of an inductive load at the drain of a power stage. A constant current pulse I_{N_1} during 50 μs with rise and falling times of 1 μs drops the voltage V_{N_1} belowground, forward biasing the N_1 well.

Figure 4.16 shows the simulation results for two different injection levels of minority carriers. In the first configuration, a current of 150 μA is injected into the substrate imposing $V_{N_1} = -0.67\,\text{V}$. This current is mainly collected by the P+ substrate contacts during the transient rise time, but after tens of μs (corresponding to the lifetime of minority carriers), the wells N_2, N_3 in the proximity are collecting most of the injected electrons. In particular N_1 collects 59%, N_2 27%, and the substrate only 14%.

In the second configuration, the injected current level is 150 mA, corresponding to $V_{N_1} = -1$ V. After 20 μs the P+ contacts collect 49% of the injected currents, while 52.5 mA flows through the nearest collector N_2 (35%) and only 24 mA to the farthest collector N_3 (16%). This means that depending on the minority carrier concentration, different situations are predicted. Therefore, in high-injection regime, the P+ contacts are more efficient to collect parasitic currents than in the low-injection case.

The GLD substrate netlist requires only 10 ms per time step to perform this transient simulation, while the simulation time needed for TCAD was around 15 s, i.e., a gain of about 1000.

4.5 Deep Trench Isolation

The substrate model can be applied to different HV technologies by appropriate meshing routines as described in Sect. 5.3.1. It is possible to explore how more expensive technology options like DTI are beneficial to address the issue of substrate currents.

Simulation of the base, emitter, and collector currents for the parasitic NPN BJT made of two 20 × 100 μm^2 n-wells is reported in Table 4.1. Low ($N_1 = -0.5$ V) and high ($N_1 = -0.8$ V) electron injection conditions are examined with N_2 biased

Table 4.1 TCAD (top line) and circuit (bottom line) results for two different belowground conditions in the five technology configurations of Fig. 4.17

−0.5 V	1	2	3	4	5
I_E	1.73 μA	1.76 μA	0.59 μA	0.13 μA	1.42 μA
	1.38 μA	1.40 μA	0.50 μA	0.10 μA	1.24 μA
I_B	0.99 μA	1.02 μA	0.15 μA	0.13 μA	0.96 μA
	0.86 μA	0.88 μA	0.14 μA	0.10 μA	0.83 μA
I_C	0.74 μA	0.74 μA	0.44 μA	5 pA	0.46 μA
	0.52 μA	0.52 μA	0.36 μA	4 pA	0.41 μA
α	0.43	0.42	0.75	4e−5	0.32
	0.38	0.37	0.72	4e−5	0.33
−0.8 V	1	2	3	4	5
I_E	2.40 mA	2.66 mA	1.84 mA	0.38 mA	0.24 mA
	2.19 mA	2.45 mA	2.00 mA	0.32 mA	0.23 mA
I_B	0.73 mA	0.99 mA	0.45 mA	0.38 mA	0.17 mA
	0.69 mA	0.98 mA	0.35 mA	0.32 mA	0.17 mA
I_C	1.67 mA	1.67 mA	1.39 mA	8 nA	0.07 mA
	1.50 mA	1.47 mA	1.65 mA	197 nA	0.06 mA
α	0.69	0.63	0.75	2e−5	0.29
	0.68	0.60	0.82	6e−4	0.26

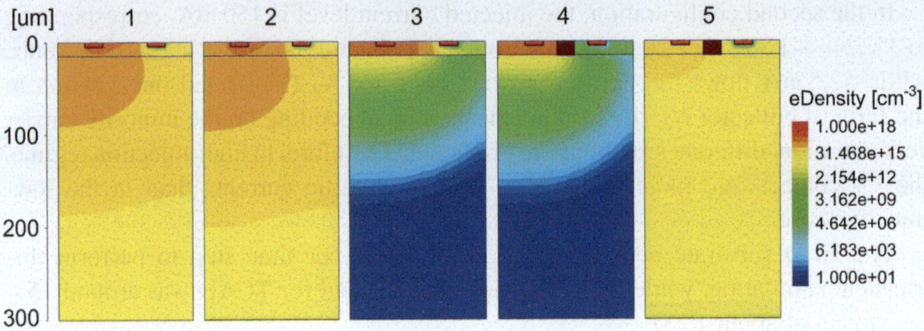

Fig. 4.17 Electron density color plot from TCAD simulations of five different structures for −0.5 V belowground condition: (1) low-doped substrate of HVCMOS, (2) low-doped substrate with backside contact, (3) P++ substrate with epi-layer and backside contact, (4) P++ substrate with epi-layer, DTI, and backside contact, (5) DTI isolation in low-doped substrate with backside contact

at 12 V. Figure 4.17 shows the electron concentration for five geometries obtained with TCAD simulations.

For simple HV-CMOS technologies (geometry 1), the two wells with an average doping of $\simeq 10^{17}\text{cm}^{-3}$ are placed inside a low-doped substrate ($\simeq 10^{15}\ \text{cm}^{-3}$) 50 μm away. This results in a coupling factor ranging from 0.4 to 0.8. In this case no backside contact is considered. In geometry 2 the backside metallization is added, showing no changes in coupling. However, for the high-injection condition, when 2.5 mA are injected into the substrate, geometry 1 reports a maximum substrate shift of 124 mV, while in geometry 2, this shift is limited to 63 mV.

Geometry 3 has been modified in order to study the effect of highly doped substrate ($\simeq\ 10^{18}\text{cm}^{-3}$) with a p-epi-layer of 15 μm thickness. In this case, both currents and couplings are strongly affected. In general lower currents are being injected, but the overall coupling between the two wells is strongly increased. This is attributed to reflection electrons experience at the PP+ interface (MCM effect). In geometry 4 a deep trench is placed that demonstrate how it forces minority carriers to recombine in the P++ layer. This technology option is effective only if the P++ substrate is present. Indeed, in geometry 5 the same DTI structure but without the highly doped substrate leads to minor improvements in current coupling issues.

In conclusion, the model can simulate different layout options and happens to be very efficient in terms of accuracy and CPU time for selecting the best solution regarding a particular circuit.

4.6 ESD Protection Examples

Several ESD protections are available for digital, analog, RF and HV ICs [12]. However, at the time being, the robustness of a protection device can only be predicted by TCAD simulations. Even if it was demonstrated to be successful in predicting the behavior of ESD clamps [7], still this requires an accurate calibration for each technology. Indeed, the technology is essential for EDS devices, and the technology-specific ESD protections forces ESD engineers to propose new protection solutions following a long and costly trial-and-error approach.

The role of ESD circuits is to clamp the voltage to safe levels and to provide a low impedance path for the current resulting from the charge transfer process. To meet these objectives, ESD designers adopt a pad-based protection scheme to drive the current sink paths through the power rails. At each I/O pad, protection circuits toward the power rails are then used.

These protections are designed to operate in a specific ESD design window. The ESD triggering voltage V_{t1} should be lower than the maximum sustainable voltage to avoid gate oxide breakdown. The triggering mechanism is usually attributed to some junction breakdown. This latching is expected to induce a snapback that clamps the voltage at a safe value V_h. This voltage is usually greater than Vdd to avoid unwanted activation of the ESD protection during regular circuit conditions. The discharge path is then characterized by an "ON" resistance R_{ON} that must be as low as possible. Finally, the robustness of the protection is defined by a second breakdown occurring at (V_{t2}, I_{t2}).

Since the second breakdown is due to thermal failures that are not included in the model, this cannot be predicted with the actual model. However, the snapback behavior can be simulated since the triggering voltage is strictly related to the avalanche breakdown of reverse-biased junctions. Some SPICE-level simulations of snapback phenomena are reported to assess the EPFL model capabilities. Moreover, geometrical and technological aspects must be taken into account during the layout of the circuit and ESD design phases. In the following section, some examples of the extended substrate model are presented where these features are illustrated with a qualitative analysis carried out for sample test structures.

4.6.1 Simulation of Breakdown in BJTs

When several PN junctions are considered, the avalanche mechanisms of simple diodes are combined with the electrical configuration, and the breakdown phenomena can be somehow amplified. The simpler case is for BJT devices where two PN junctions with different breakdown properties are present.

Connecting together two lumped diodes, the BJT breakdown effect can be simulated without additional parameters. This process can then be further extended to more complex devices such as thyristors. To show the validity of this approach, the simple NPN structure with two identical junctions is simulated. Here, 200 nm highly doped N+, P+ contacts are added with the corresponding homojunction contact models that strongly influence the gain of the transistor. Typical high-voltage technology doping concentrations are used to simulate the breakdown mechanisms of a lateral bipolar transistor between the two n-wells in a low-doped substrate kept at ground.

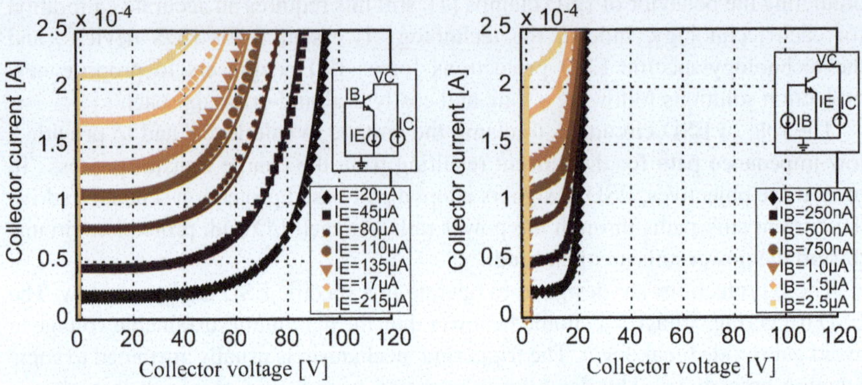

Fig. 4.18 (Left) Common-base (CB) configuration and breakdown simulations. (Right) Common-emitter (CE) configuration and breakdown simulations. Continuous lines refer to circuit model, the points to TCAD simulations

Several attempts [10] have been done to include breakdown in compact models as an additional diode sinking a breakdown current. However, this approach works only for diodes, while for BJT it is mandatory to consider breakdown currents in terms of a multiplication factor and not as an additional factor. Indeed, internal couplings create feedback changing the breakdown mechanism. This depends on the electrical configuration of the three terminals of the transistor. To this purpose, Common base (CB) and Common emitter (CE) configurations are simulated for different emitter and base currents and are shown in Fig. 4.18.

For identical junctions, the breakdown voltage and the transport factor are $BV = 74$ and $\alpha = 3.9$. Without additional parameters other than the doping and lengths, the variation of the early effect (present only in the CE case) and the shift of the breakdown voltage from $BV_{\mathrm{CB}} = 96\,\mathrm{V}$ to $BV_{\mathrm{CE}} = 24\,\mathrm{V}$ are consistent with TCAD simulations. The model predicts internal feedbacks and shifts in the breakdown voltage depending on the electrical configurations.

To understand these results, a basic circuit analysis of the BJT is proposed. The emitter and collector currents are linked by the transport factor α as:

$$I_C = -\alpha I_E M \tag{4.1}$$

In the CB configuration, the emitter current is constant, and the collector current starts to increase only when the base-collector junction breaks down. In the CE case, Eq. (4.1) is still valid, but I_E is no longer constant.

$$I_E = \frac{I_B}{1 - \alpha M} \tag{4.2}$$

Then, given a fixed I_B, the collector current is influenced by the current gain of the BJT, as shown in Eq. (4.3).

$$I_C = \alpha I_B \frac{M}{1 - \alpha M} \tag{4.3}$$

In particular, the breakdown voltage in the common emitter configuration is strongly reduced by internal feedback mechanisms, as shown by the simulations.

4.6.2 Diode Chains Leakages

Diodes are the simplest and widely used ESD protection devices. If in reverse bias, the avalanche breakdown is the triggering mechanism for the ESD clamping, in forward bias they naturally switch on around the built-in voltage V_{bi}. In real circuits, these diodes are usually connected in reverse bias between the I/O pad and the power rails. However, apart from N+ PSUB diode used between the I/O pad and ground, other diode implementations are realized to separate the anode from the common substrate. Substrate-isolated integrated diodes are then created between isolated P+ and n-well regions. However, a parasitic vertical PNP transistor is present between the n-well region and the substrate.

Simulations of simple diode structures have been presented already. Here the EPFL substrate model is used to simulate the effects of the parasitic BJT leakages in ESD structures involving multiple diodes. This is a typical case for power clamping circuits based on the diode chain where multiple junctions are connected in series in order to increase the holding voltage to values higher than the Vdd. Since each single diode is forward biased around 0.7 V, the diode chain clamping is usually restricted to low-voltage applications.

Two examples of a diode chain implementation with the corresponding parasitics are shown in Figs. 4.19 and 4.20. The first configuration involves several diodes in series in forward bias and where the vertical PNP BJT is leaking into the substrate. During the activation of the diodes, the current is expected to propagate mainly from the emitter p-well to the substrate PSUB (high-leakage path). The second implementation is a low-leakage diode chain as it relies on isolated diodes [3]. The diode between the p-well and N+ is isolated by an n-well connected to the local anode (highest potential). This connection ensures that the base-emitter junction of the parasitic PNP BJT is shorted and does not inject holes into the substrate. However in this case, a parasitic NPN BJT is also present.

The first diode chain corresponds to a Darlington configuration of the parasitic PNP BJTs, as reported in the literature [6]. This connection has several consequences. First the overall current collected by the chain cathode is decreasing with the number of devices N. In fact this current corresponds to the base current of the last device $I_{B,N}$, while the chain anode current is the emitter current of the first device $I_{E,1}$. Since part of the current flowing through the chain is lost into the substrate, the last base current is affected by the gain of each stage:

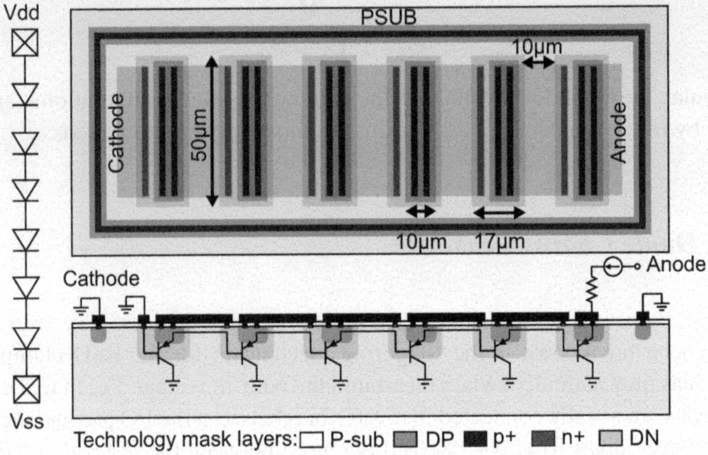

Fig. 4.19 Layout view (top) and cross section of the parasitic BJTs (bottom) for the high-leakage diode chain (drawings not to scale)

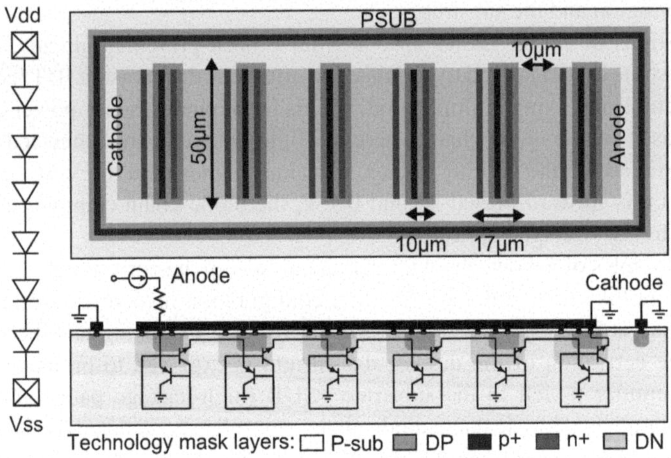

Fig. 4.20 Layout view (top) and cross section of the parasitic BJTs (bottom) for the low-leakage diode chain (drawings not to scale)

$$I_{B,N} = I_{E,N-1} = \frac{I_{E,1}}{\Pi_{k=1}^{N}(1+\beta_k)} = \frac{I_{E,1}}{(1+\beta)^N} \tag{4.4}$$

Therefore, the substrate current is almost equivalent to the total current. These leakage currents change slightly with the number of diodes as shown in Eq. (4.5).

$$I_{\mathrm{SUB}} = \beta I_{E,1} \sum_{k=1}^{N} \frac{1}{(1+\beta)^k} = I_{E,1} \frac{(1+\beta)^N - 1}{(1+\beta)^N} \simeq I_E \tag{4.5}$$

Combining the two equations, the β boosting parameter of the Darlington configuration is derived. For N stages, the equivalent β_{eq} between the common collector (i.e., the substrate) and the last base (i.e., the chain cathode) is increasing as a power law (see Eq. (4.6)). The more diodes are added to the chain, the lesser the current is flowing from the anode to the cathode.

$$\beta_{eq} = \frac{I_{\mathrm{SUB}}}{I_{B1}} = (1+\beta)^N - 1 \simeq \beta^N \tag{4.6}$$

Finally, the total voltage drop in forward bias is not proportional to the number of diodes. The V_{BE} of the N-th device is affected by the gain of the previous one in a logarithmic dependence:

$$V_{\mathrm{BE},N} = nV_t \ln \frac{I_{E,N}}{I_S} = nV_t \ln \frac{I_{E,N-1}}{(1+\beta)I_S} = V_{\mathrm{BE},N-1} - nV_t \ln(1+\beta) \tag{4.7}$$

Since the final trigger voltage is the sum of all the V_{BE}s along the chain, if N diodes are present, the triggering voltage is:

$$V_{t1} = \sum_{k=1}^{N} V_{\mathrm{BE},k} = NV_{bi} - \frac{N(N-1)}{2} nV_t \ln(1+\beta) \tag{4.8}$$

where the second term takes into account the Darlington amplification effect. So, for instance, when $V_{\mathrm{DD}} = 3.3\,\mathrm{V}$, $N = 6$, diodes should be used to target $V_{t1} \simeq 4.2\,\mathrm{V}$ if $V_{bi} = 0.7$ and $n \simeq 1$. However, due to the correction factor, almost 1 V is lost when $\beta \simeq 10$. Therefore, for high voltages, long diode chains are required, but still these will not be efficient.

For the low-leakage diode chain with isolated diodes as in Fig. 4.20, the analysis is simpler as the PNP BJTs is not activated, and the total current flow from the anode to the cathode of the chain, mainly passing through the N isolation well (which represents the collector of the diode-connected parasitic NPN BJTs). In this case, current leakage into the substrate is low and does not vary with the current flowing into the structure.

All those effects have been simulated by varying the number of diodes in a chain. The I–V results of three-dimensional simulations are reported in Fig. 4.21. As expected, while for the low-leakage chain, the triggering voltage is increasing linearly with the number of devices, the high-leakage chain degrades due to the Darlington effect. In Fig. 4.22 the β of this diode string is reported and shows that the Darlington multiplication factor of the parasitic vertical PNP BJT is $\beta \simeq 8$.

The simulated parasitic substrate current as a function of the current flowing into the chain is compared in Fig. 4.23 for the two topologies. As expected, in the

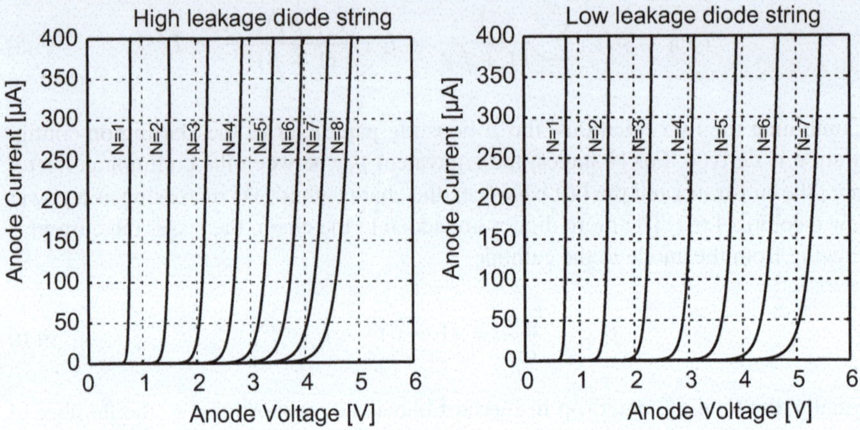

Fig. 4.21 Simulated *IV* curves for the high-leakage (left) and low-leakage (right) diode strings

Fig. 4.22 Simulated β multiplication for the Darlington connected diode chain

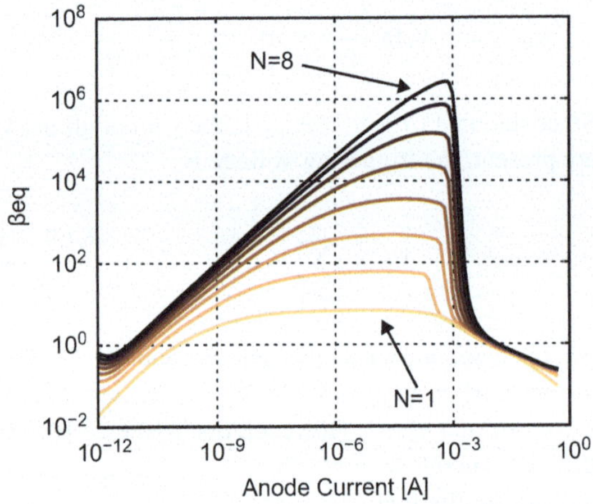

first case, the leakage is proportional to the current flowing into the structure, while the cathode current is decreasing when more diodes are added. Instead, in the low-leakage chain, the leakage current remains around 300 fA for a wide range of current densities. These results demonstrate that the substrate model can quantify current leakages for a given layout and for a given technology.

Unexpected behaviors are simulated for high currents. For the high-leakage diode chain, the cathode current is increasing up to 1 mA, while the leakage currents of the low-leakage chain are ramping up to approximately 33 mA. Most likely, this is attributed to the activation of parasitic components. The extended *IV* curves reported in Fig. 4.24 confirm that a snapback takes place in the high-leakage diode chain.

To understand what is the parasitic device responsible for the snapback, simulated voltages and currents are reported in Fig. 4.25 for $N = 8$ before ($I_A = 100\,\mu\text{A}$)

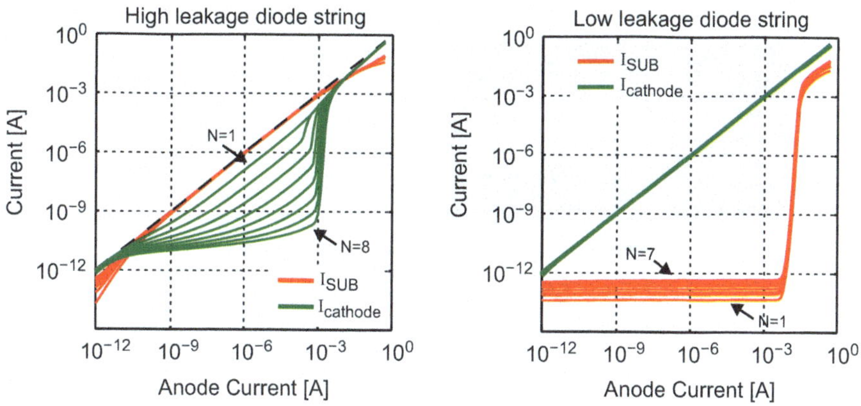

Fig. 4.23 Simulated cathode and substrate currents (i.e., leakages) for the two diode chain structures

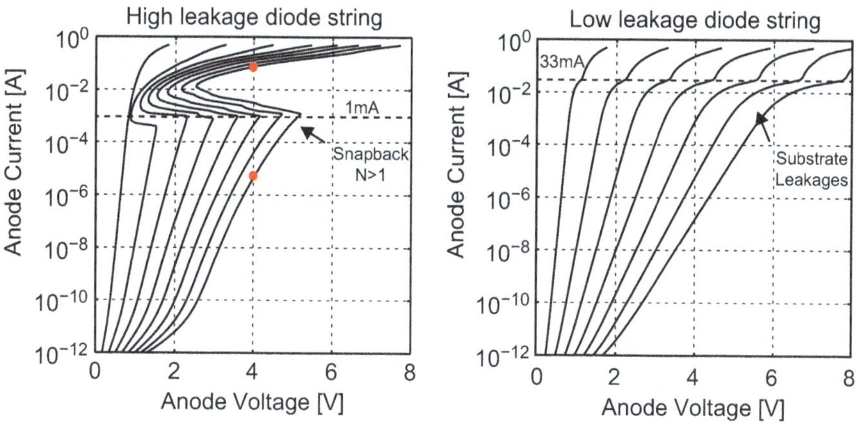

Fig. 4.24 Extended *IV* curves simulations of the diode strings up to 0.5 A

and after ($I_A = 1$ mA) snapback. Before snapback the voltage drops linearly across the n-wells with an emitter current decreasing from the anode to the cathode. The current flows primarily into the substrate through the vertical PNP BJTs in a nonuniform manner that depends on the layout configuration. After snapback, a large positive substrate voltage shift is recorded, and the substrate current decreases. Looking at the voltage distribution, all the sides of the well junctions are forward biased. They create a path through the parasitic NPN BJTs and increase the current in the last cathode. The activation of the lateral transistors takes place when $N \geq 2$ and is induced by the leakage currents. In fact these currents passing through the substrate shift upward the local potential and forward bias the wells. When more than one diode is in the chain, the last NPN parasitic BJT will be always turned on because the last n-well is at a low potential (due to the voltage drop across

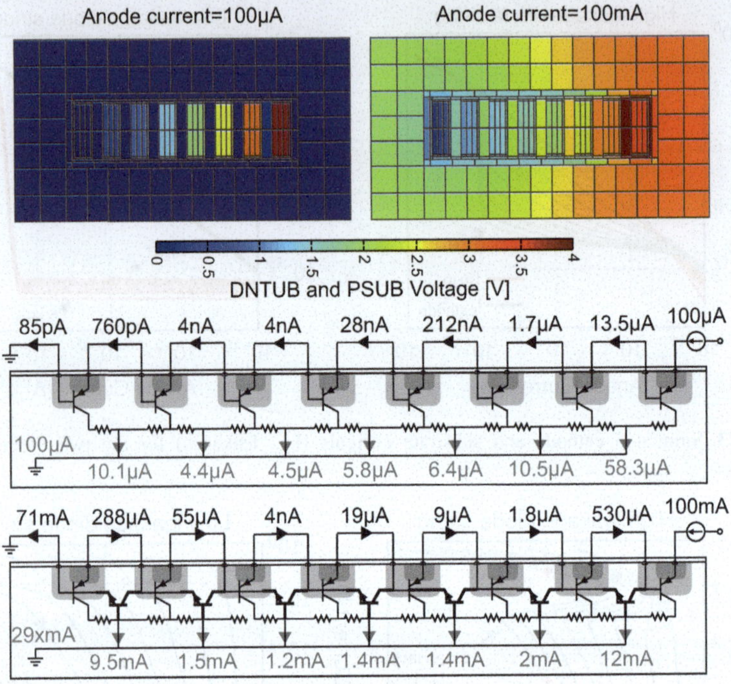

Fig. 4.25 Simulated voltage (top) and current (bottom) of the high-leakage eight-diode string before and after snapback

the devices). To check what are the parasitic NPN BJTs that are forward biased, the inspection of the minority carriers injected is reported in Fig. 4.26. Due to the symmetry of the layout, it is clear that the first and last BJTs are the most forward biased and inject electrons into the substrate.

For the low-leakage diode string configuration, the situation is different, and no snapback is reported, meaning that no activation of lateral NPN BJT is evidenced. In this case, the current is forced to flow inside the n-isolated wells. However, for high currents a local potential shift due to the finite R_{well} forward biases the parasitic vertical PNP BJT and injects leakage currents into the substrate. To confirm this hypothesis, Fig. 4.26 plots the simulated voltage distribution in the n-wells and in the substrate where a gradient is clearly visible in the wells. The same color plot shows that the lateral NPN BJTs are however not activated in this configuration.

4.6.3 Transistor-Based Protections

BJTs can be used as simple snapback devices if the collector is connected to the I/O pad and a base bias V_B is provided [1]. During normal circuit operation, the

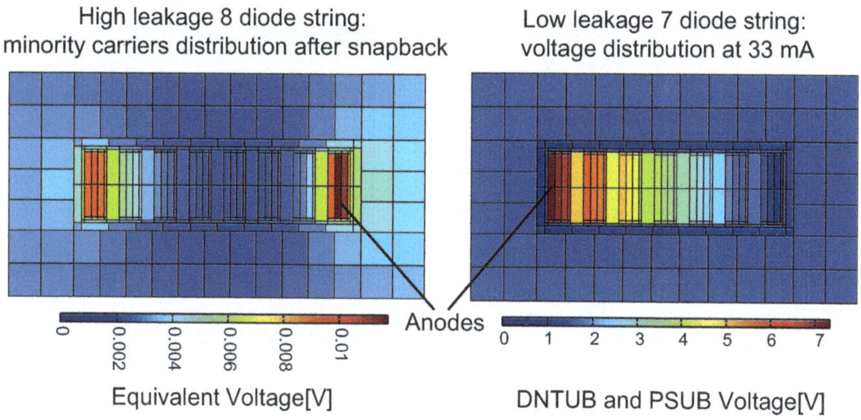

Fig. 4.26 Inspection of simulated injected electrons distribution after snapback of the high-leakage eight-diode string (left) and of the potential shift of low-leakage seven-diode string (right)

transistor is off if V_B is chosen to ensure the reverse biasing of emitter and collector junctions (e.g., grounding the base). When a positive ESD pulse appears on the collector, the collector voltage rises up until it reaches the breakdown voltage of the collector junction. At this point, a current starts to flow toward the base, rising the base potential and forward biasing the base-emitter junction. A discharge channel is then activated through the transistor, and the collector voltage snaps back to a lower holding voltage since the transistor is turned on. This results in a pad voltage clamping at a holding voltage V_h.

The EPFL substrate model can simulate the snapback for the typical configuration of a grounded base BJT. The structure reported in Fig. 4.27 is simulated by varying the value of the base resistance. This snapback protection is influenced by the base resistance R_b and is often used to control the holding voltage of the ESD clamp as an external element. Without adding any new component in the model, Fig. 4.28 shows the results for the grounded base NPN BJT. IV simulations with SPICE software are obtained in a few seconds, and some well-known conclusions about BJT protection design can be derived. The triggering voltage V_{t1} of the BJT depends on the collector breakdown voltage BV_{CBO} , while the holding voltage V_h is related to the geometry and to the external resistor. Therefore, the substrate model can be used to predict V_h once the layout and/or circuit modifications are done. Moreover, in the same plot, the absolute value of the ESD current for negative ESD pulse is also shown. In this case, the base-collector junction is directly forward biased shunting the current to ground without any snapback.

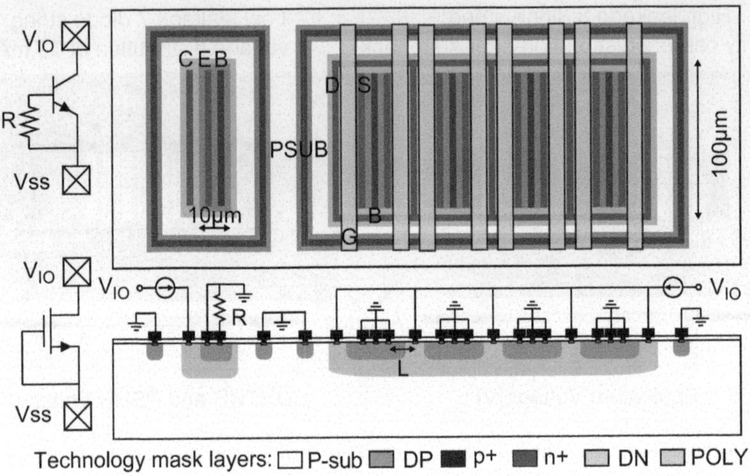

Fig. 4.27 Layout view (top) and cross section (bottom) for the simulated grounded base BJT (left) and ggMOS (right). Drawings are not to scale

The parasitic NPN BJT in MOS structures is widely used to implement ESD protections for standard 2 kV Human body model (HBM) specifications (see Fig. 4.27). In this application, a fingered transistor is usually laid out with finger widths around 50–100 µm to achieve a uniform current distribution. For this device the gate is grounded (ggNMOS) to ensure zero leakage during normal IC operation.

In the case of an LDMOS, the drain D holds for the equivalent collector, the source S for the equivalent emitter, and the bulk B for the equivalent base. Then, the activation of the ggMOS can be easily interpreted. At positive ESD events, the reverse-biased DB junction breaks down, and holes are injected into the bulk, rising the potential across the p-well parasitic resistance. The SB junction is then forward biased letting the current pass through the parasitic NPN BJT. In this case, a snapback is also expected. If instead a negative ESD event appears, the DB junction is forward biased and shunts the current.

Circuit simulations are reported in Fig. 4.28, confirming the ggMOS snapback behavior. Three main layout parameters are involved in ggNMOS design: the width related to the maximum current, the length L that corresponds to the base length of the parasitic bipolar transistor, and the gate-to-drain spacing that introduces a ballasting resistor in the collector for uniform triggering of a multi-fingered structure. Simulations show that it is possible to track the holding voltage variations with respect to the gate length by using the EPFL substrate model. Particularly for a long base length L, the gain of the NPN BJT is lowered, and the holding voltage is increased [14].

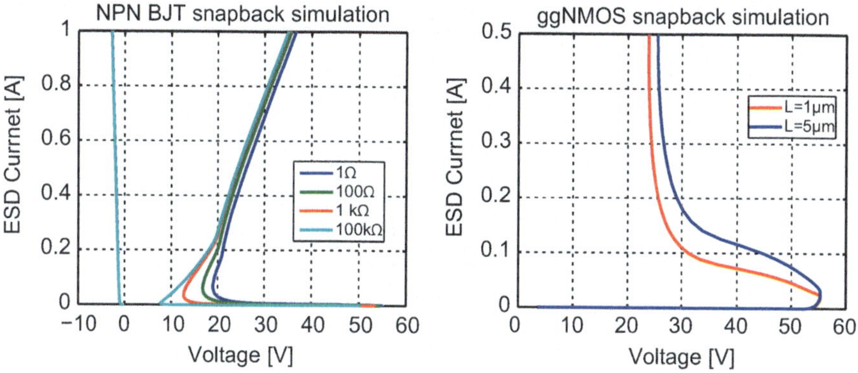

Fig. 4.28 Simulated *IV* curves showing snapback of the NPN BJT (left) and ggMOS (right)

4.6.4 Silicon-Controlled Rectifiers (SCR)

Another widely studied class of ESD protection is based on Silicon-controlled rectifier (SCR) composed of more than two junctions, i.e., a thyristor-like structure. When SCRs latch, they provide a low-current discharging path for the ESD charges, making these rectifiers the most robust ESD solution (for the same silicon area). However, the presence on an internal positive feedback that sustains the current discharge can also damage the chip if the protection is triggered during normal IC operation.

Two examples of lateral SCR structures are compared by using the proper extended GLD [2]. Also SCRs are in fact snapback devices where the avalanche of one or more junctions is the main triggering mechanism. The two selected structures reported in Fig. 4.29 are the one-direction SCR and the dual-direction SCR. As the name suggests, the first one has a snapback only for positive anode voltages, while the other has a double snapback that extends its use to bidirectional ESD clamping.

The substrate model SPICE simulations are reported in Fig. 4.30 where the difference among these architectures is clearly visible. The asymmetry can be understood from the cross-section representation of the devices. The one-direction SCR is a PNPN structure with three junctions, while the dual-direction SCR is a NPNPN structure with four junctions. Devices made of an odd number of junctions are unidirectional (e.g., the diode), while those consisting in an even number of junctions are bidirectional (e.g., the BJT). In the latter case, the two propagation directions have different characteristics depending on the geometry and the technology.

Simulated results show how the EPFL substrate model can simulate SCRs structures allowing the designer to check unidirectional or bidirectional snapback current paths.

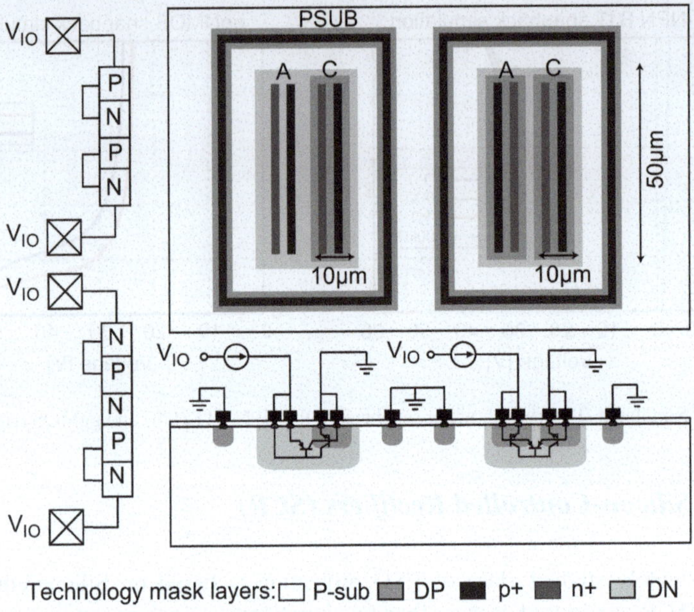

Fig. 4.29 Layout view (top) and cross section (bottom) for the simulated grounded SCR (left) and dual-direction SCR (right). Drawings are not to scale

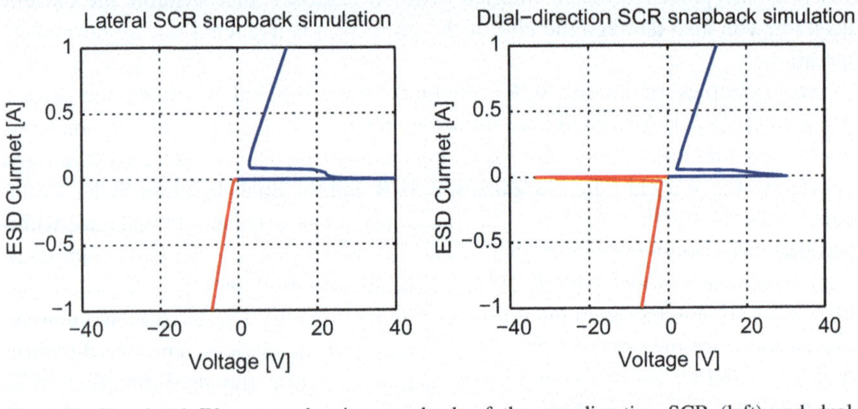

Fig. 4.30 Simulated *IV* curves showing snapback of the one-direction SCR (left) and dual-direction SCR (right) of Fig. 4.29

4.7 Conclusions

The GLD was coded in VerilogA, and extensive simulations were performed to validate the EPFL substrate model. The results were compared with TCAD simulations considered as the most accurate reference. The same technology parameters and temperature dependencies were used. DC, AC, and transient simulations are well predicted for diodes, bipolar transistors, as well as parasitic coupling mechanisms between the different wells in multi-collector configurations.

High-injection effects and conductivity modulation in diodes are also taken into account with the EPFL diode model. The impact of doping discontinuities on the minority carrier transport is discussed for different structures and can be addressed with the EPFL homojunction model. Finally the EPFL resistor model for the substrate takes into account the geometry dependencies of the current gain of parasitic lateral NPN BJTs as well as the potential shift after the activation of vertical PNP BJTs. Examples illustrating how the model can be adapted to different HV technologies are also discussed. In particular, the deep trench isolation option can be simulated, and the model predicts a drastic decrease in the lateral coupling arising from minority carriers.

Finally, breakdown simulations for ESD devices are reported. The set of lumped devices interconnected with a proper mesh can track the unstable breakdown phenomena in different BJT configurations and so can be used to simulate snapback behavior in ggMOS and SCR and detect the activation of parasitic substrate devices, such as in diode chains, for instance.

Compared to TCAD, the GLD extracted substrate netlist gives a gain of one thousand in terms of simulation time. The EPFL substrate model is then a promising solution for designers to investigate substrate couplings in ICs starting from the layout.

References

1. G. Bertrand, C. Delage, M. Bafleur, N. Nolhier, J.-M. Dorkel, Q. Nguyen, N. Mauran, D. Tremouilles, P. Perdu, Analysis and compact modeling of a vertical grounded-base n-p-n bipolar transistor used as ESD protection in a smart power technology. IEEE J. Solid-State Circuits **36**(9), 1373–1381 (2001)
2. M.-D. Ker, K.-C. Hsu, Overview of on-chip electrostatic discharge protection design with SCR-based devices in CMOS integrated circuits. IEEE Trans. Device Mater. Reliab. **5**(2), 235–249 (2005)
3. M.-D. Ker, W.-L. Wu, ESD protection design with the low-leakage-current diode string for RF circuits in BiCMOS SiGe process, in *Electrical Overstress/Electrostatic Discharge Symposium (EOS/ESD)* (2005), pp. 1–7
4. I. Ladany, An analysis of inertial inductance in a junction diode. IRE Trans. Electron Devices **7**(4), 303–310 (1960)
5. S.E. Laux, K. Hess, Revisiting the analytic theory of p-n junction impedance: improvements guided by computer simulation leading to a new equivalent circuit. IEEE Trans. Electron Devices **46**(2), 396–412 (1999)
6. T.J. Maloney, S. Dabral, Novel clamp circuits for IC power supply protection, in *Electrical Overstress/Electrostatic Discharge Symposium (EOS/ESD)* (1995), pp. 1–12
7. J.A. Salcedo, J.J. Liou, Z. Liu, J.E. Vinson, TCAD methodology for design of SCR devices for Electrostatic Discharge (ESD) applications. IEEE Trans. Electron Devices **54**(4), 822–832 (2007)
8. M. Schenkel, Substrate current effects in smart power ICs, PhD thesis, ETH Zürich, Nr. 14925, 2003

9. E. Seebacher, W. Posch, K. Molnar, Z. Huszka, Analog compact modeling for a 20–120 V HV CMOS Technology, in *Proceedings of NSTI Nanotechology Conference Trade Show*, vol. 3, pp. 720–723 (Nano Science and Technology Institute, Amritsar, 2006)
10. Y. Subramanian, R.B. Darling, Compact modeling of Avalanche breakdown in pn-junctions for computer-aided ESD design (CAD for ESD), in *Technical Proceedings of the International Conference on Modeling and Simulation of Microsystems* (2001), pp. 205–208
11. J.J.H. van den Biesen, Modelling the inductive behaviour of short-base p-n junction diodes at high forward bias. Solid State Electron. **33**(11), 1471–1476 (1990)
12. V.A. Vashchenko, A. Shibkov, *ESD Design for Analog Circuits* (Springer, New York, 2010)
13. J.-S. Yuan, J.J. Liou, W.R. Eisenstadt, A physics-based current-dependent base resistance mode for advanced bipolar transistors. IEEE Trans. Electron Devices **35**(7), 1055–1062 (1988)
14. J. Yuxi, L. Jiao, R. Feng, C. Jialin, Y. Dianxiong, Influence of layout parameters on snapback characteristic for a gate-grounded NMOS device in 0.13-μm silicide CMOS technology. J. Semicond. **30**(8), 084007 (2009)

Chapter 5
Extraction Tool for the Substrate Network

5.1 EPFL Substrate Model Extraction Flow

A view of the substrate extraction process is shown in the flow diagram of Fig. 5.1. The IC layout is divided into a number of three-dimensional cells to extract the model geometrical parameters, such as length and area of each mesh unit. Specific layout layers correspond to specific technology parameters, such as substrate material type, doping, and well depth. This technology information is saved in two files resulting from a technology reduction process.

Based on these files, the IC layout in connection with the mesh [2] is scanned to extract the substrate model during the LVS verification. For this purpose, the technology extract rule file was modified to extract the EPFL substrate parasitic diodes, homojunctions, and diffusion resistors with their geometrical parameters and connectivity obtained by the LVS tool. The connection between the meshing nodes is done by means of EPFL diodes and homojunctions. The connection between nodes of the same doping is done via EPFL diffusion resistors. The extracted substrate model network automatically back-annotates bipolar parasitic transistors to the circuit devices. In the flow diagram of Fig. 5.1, each rectangle represents a complex process detailed in this chapter. These processes were integrated in the mixed-mode design flow based on commercial EDA tools.

5.2 Technology Reduction

To build the substrate equivalent circuit model, the IC layout is divided into mesh elements in order to apply the FDM scheme. Since the extracted network consists of a large number of cells that may result in large simulation times, a layout simplification is applied to the IC layout before generating the mesh.

P. Buccella et al., *Parasitic Substrate Coupling in High Voltage Integrated Circuits*, Analog Circuits and Signal Processing, https://doi.org/10.1007/978-3-319-74382-0_5

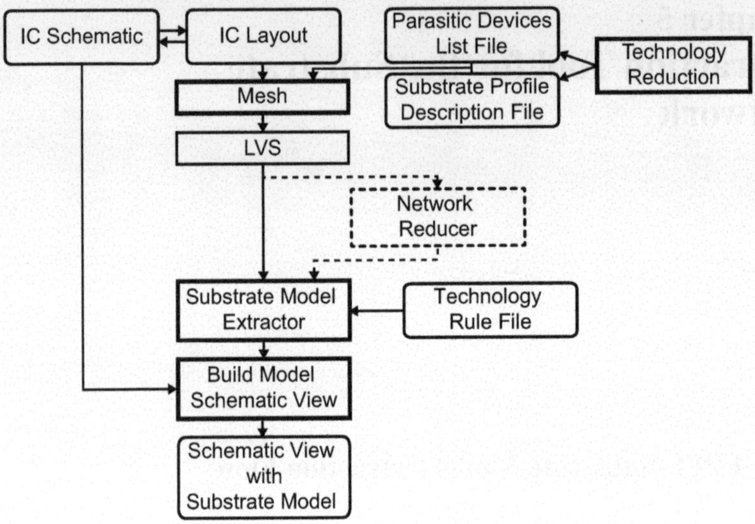

Fig. 5.1 Flow diagram with the key steps for the extraction of the substrate model

This relies on two steps:

1. *Parasitic device identification*: first, all possible combinations of parasitic PN junctions and homojunctions are identified for a given technology. This step is done by a routine based on geometrical Boolean functions checking the spatial relationships between layer geometries.
2. *Parasitic device reduction*: next, the parasitic devices identified are selected and classified according to the circuit application and robustness requirements. The most critical parasitic devices are selected to build the model of the substrate.

This reduction process can be applied to any technology such as HVCOMOS, modern BCD, and SOI.

5.2.1 Parasitic Device Identification

When drawing the layout of power devices like NDMOS, PDMOS, BJTs, and digital NMOS and PMOS, a large number of layout layers are required such as n-wells and p-wells, polygate, metals, etc. . . specified by the technology process flow. Nevertheless, only some of the mask layers are relevant to identify substrate parasitic devices.

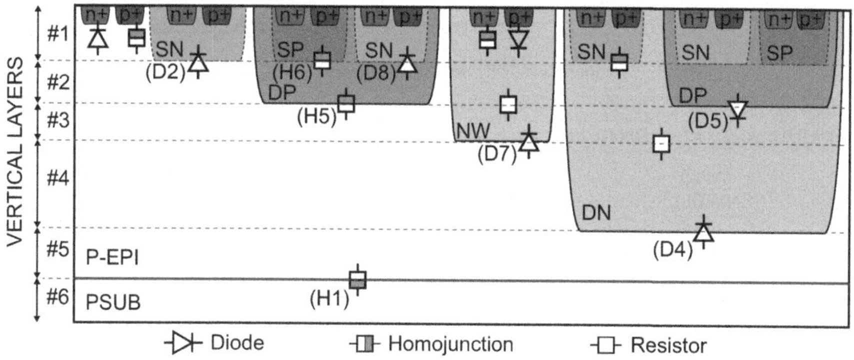

Fig. 5.2 Cross section of a high-voltage process. The substrate is treated as a multilayer structure in the vertical direction where each well is assumed to have a constant doping concentration

As explained in Chap. 5, the substrate model is a network that consists of three electrical elements: PN junctions, homojunctions (at doping discontinuities between layers of same type), and diffusion resistors. Thus, p- and n-type regions provide enough information to identify critical layers and extract the substrate model. These layers have been investigated in for a HV triple-well CMOS technology. The process technology is based on the so-called Smart voltage extension (SVE) approach [1]. To create high-voltage circuit elements, additional shallow and deep n-wells (SN and DN) as well as shallow and deep p-wells (SP and DP) are implanted into the p-type substrate [9]. A p-type epitaxial layer is often available as a process option and is accounted in the parasitic device identification. To identify parasitic devices, a layout including all available circuit devices (excluding poly and metal layers) is drawn as shown in Fig. 5.2. In this way all process layers are visible. In the vertical direction, the substrate is treated as a multilayer structure where each well is assumed to have a constant averaged doping concentration.

To identify PN junctions and homojunctions, a hierarchical tree diagram is used to represent the technology as a nested set of doping regions. Figure 5.3a is a tree structure showing the hierarchical organization of the technology with respect to the doping level and depth while considering the substrate as the root. The tree diagram nodes are the wells and the connecting lines define the relationship between them. Lines connecting layers of the same doping type indicate an homojunction, while lines connecting layers of different doping type indicate PN junctions. The homojunctions between the external and the internal nodes on the tree define the substrate and the well biasing contacts. A visual inspection of the tree reveals nine different PN junctions and height homojunctions (without duplication).

This representation of the layout as a tree shows all possible electrical parasitic paths in a given technology. Indeed, simple paths that include either front-to-front or two back-to-back connected diodes describe lateral or vertical bipolar devices (NPN or PNP) [3], while other complex paths identify parasitic structure such as a thyristor (PNPN). The list with all parasitic devices after reduction is stored in a

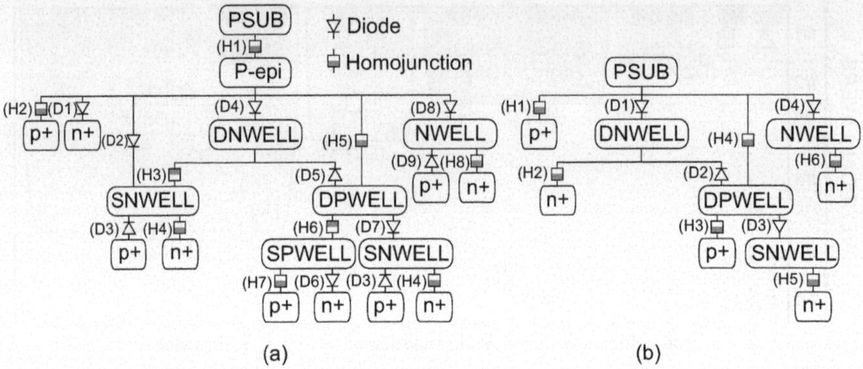

Fig. 5.3 (**a**) Hierarchical layout view relationships for parasitic device identification. (**b**) Hierarchical layout view after parasitic device reduction

parasitic device list file that also includes the model parameters (i.e., doping and mobility as required for simulation).

5.2.2 Parasitic Device Reduction

To simulate substrate parasitic couplings, it is not necessary to extract all components as shown in Fig. 5.3a. Simplification rules based on the process and on critical circuit aspects are required to reduce the number of components. This important step allows to simplify the substrate network while keeping the most relevant information for the parasitic coupling analysis.

Initial simplifications consist in removing some process layers. Generally, homojunctions with a high doping discontinuity must be extracted [11], such as for contacts to the substrate and to the wells (i.e., p+ and n+ contacts), while homojunctions with a low doping discontinuity can be replaced by an averaged doping (i.e., SP+DP $\Rightarrow$ DP region). This step is done by applying geometrical functions to the layout that remove unnecessary layers. For example, when DN and SN or DP and SP layers overlap each other, they are merged, and only DN and DP layers are retained. Moreover, the deepest well layer is retained for the vertical meshing. In Fig. 5.4a, the cross of a NDMOS is shown with its layers. After parasitic device reduction, layers SP and DP, SN, and DN are merged, and the number of vertical layers is reduced as well.

Further layout simplification is derived with respect to the applications and electrical robustness requirements. In HV technologies, the high-voltage DN region represents either the drift region of a NLDMOS (Fig. 5.4a) or the body of the complementary PLDMOS transistor. In many applications, the high-voltage body region of the NLDMOS is connected to the corresponding source contact by a metal as shown in Fig. 5.4b. Parasitic devices are further reduced by removing short-circuited

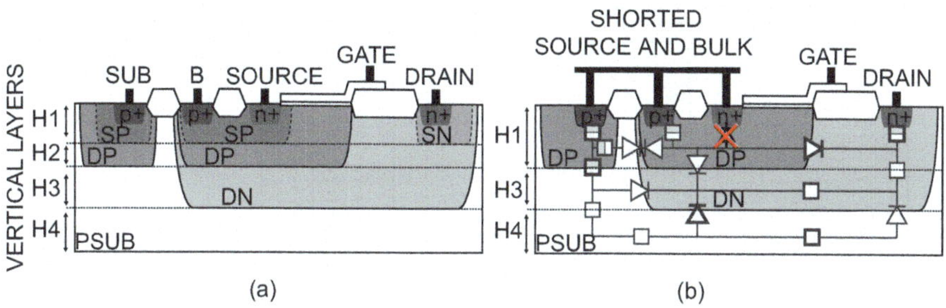

Fig. 5.4 (**a**) NDMOS cross section with process layers. (**b**) NDMOS-simplified cross section with equivalent substrate model after parasitic device reduction

PN junctions such as for the DP-n+ diode in Fig. 5.4b. This simplification must be avoided in a high-voltage PLDMOS switch with a built-in reverse voltage protection circuit [6]. Indeed, in that case, PLDMOS source-bulk and drain-bulk diodes could be forward biased under certain operating conditions and affect substrate couplings. This aspect is described in Sect. 6.3.4 dedicated to the investigation of an integrated rotor coil driver where a latch-up is triggered by the activation of the source-bulk diode.

In our case, following the reduction steps discussed previously, PSUB, DP, and DN layers are considered for the extraction of parasitic substrate elements (Fig. 5.4b). Note that n+ and p+ doping layers are excluded from this reduction process as they represent the connections to the source or drain of active devices.

Once these simplifications are done, the hierarchical layout shown in Fig. 5.3b contains only useful parasitic devices. The number of diodes to be extracted was reduced from nine to four, the number of homojunctions from eight to six, and the number of diffusion resistors from eight to five. The substrate profile description file contains the thicknesses and the doping levels of the selected layers.

5.3 Substrate Mesh Generation

The main drawback is the large number of nodes generated by the substrate model that also increases with the size of the layout. Some techniques must be developed to reduce the density of nodes. As a matter of fact, compared to the FEM, the FDM suffers from the same issue, especially when geometry misalignments are present. Up to now, the meshing grid is composed of aligned points where the key meshing lines (e.g., the border of diffusion wells) are propagated all along the substrate to create the meshing cells. An example of such a grid is shown in Fig. 5.5a. Boundary lines of each p- and n-type layout layers are extended on the surface along x and y directions resulting in a rectilinear grid covering the whole substrate.

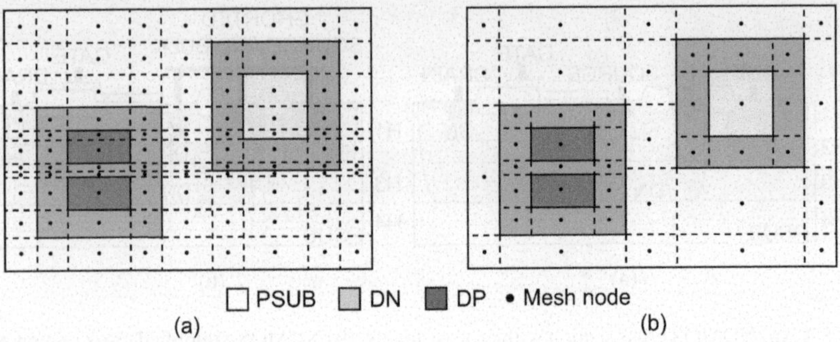

Fig. 5.5 Layout uniform meshing strategy (**a**) and proposed misaligned mesh strategy (**b**)

By default, such a grid is applied to the entire IC layout and generates a large number of nodes. These can be reduced with special node reduction algorithms. However, elimination of nodes depends on various rules that may also require long computations. The simple strategy based on merging nodes of the same doping into a single node requires, however, many specific rules to avoid information loss. For instance, merging nodes belonging to the p-type substrate must be done with great care to avoid non-orthogonal routing and layout misalignment.

5.3.1 Layout Misalignment and Finite Box Method

In order to reduce the number of nodes, some misalignment inside the grid can be tolerated. This problem can be solved by the FBM [4]. To understand how the FDM can be adapted to non-orthogonal layout, we consider the 2D configuration reported in Fig. 5.6 where the node $(i, j+1)$ is missing. This is the simplest case of a mesh discontinuity where one meshing node (i, j) must be connected to two meshing nodes along one direction. From the central difference scheme, a particular importance is given to the derivative at the midpoint $(i+\frac{1}{2}, j)$ falling between three cells. The FBM approach is based on the assumption that the contribution of $\frac{\partial u}{\partial x}\big|_{i+\frac{1}{2},j}$ can be modeled as a linear interpolation of the gradient in the virtual nodes $(i+\frac{1}{2}, j+1)$, $(i+\frac{1}{2}, j-1)$. The weighting coefficients are the shared areas A_1, A_2 between the neighbor cells, as reported in Eq. (5.1a). Using the central difference scheme, the gradient along the path $x \in [x_{i,j}, x_{i,j\pm1}]$ is split in a contribution from the gradient along $x \in [x_{i,j}, x_{i+1,j+1}]$ and $x \in [x_{i,j}, x_{i+1,j-1}]$ (see Eq. (5.1b)).

$$A \left.\frac{\partial u}{\partial x}\right|_{i+\frac{1}{2},j} = A \frac{A_1 \left.\frac{\partial u}{\partial x}\right|_{i+\frac{1}{2},j+1} + A_2 \left.\frac{\partial u}{\partial x}\right|_{i+\frac{1}{2},j-1}}{A} \tag{5.1a}$$

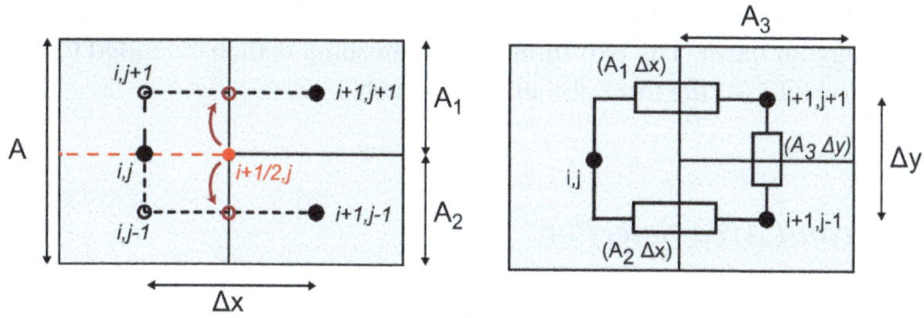

Fig. 5.6 Meshing cell misalignment analysis with equivalent circuit connections

$$A\left.\frac{\partial u}{\partial x}\right|_{i+\frac{1}{2},j} = A_1\left(\frac{u_{i+1,j+1}-u_{i,j+1}}{\Delta x}\right) + A_2\left(\frac{u_{i+1,j-1}-u_{i,j-1}}{\Delta x}\right) \tag{5.1b}$$

The quantities $u_{i,j\pm 1}$ are related to $u_{i,j}$ by a linear interpolation as well,

$$u_{i,j} = \frac{A_1 u_{i,j+1} + A_2 u_{i,j-1}}{A_1 + A_2} \tag{5.2}$$

Using this expression in Eq. (5.1b) gives Eq. (5.3) with weighted areas:

$$A\left.\frac{\partial u}{\partial x}\right|_{i+\frac{1}{2},j} = A_1\left(\frac{u_{i+1,j+1}-u_{i,j}}{\Delta x}\right) + A_2\left(\frac{u_{i+1,j-1}-u_{i,j}}{\Delta x}\right) \tag{5.3}$$

In Fig. 5.6 the equivalent circuit is shown where the area and the length of each component are properly computed by the meshing algorithm. The approximation on the gradient can be interpreted as follows: the spatial variation of the quantity $u_{i,j}$ in the normal direction Δy is neglected, i.e., $u_{i,j} \simeq u_{i,j+1} \simeq u_{i,j-1}$. This error is bigger for larger misalignment. Additional studies should be done to quantify it.

The FBM method reduces the number of nodes compared to the rectilinear solution. Based on the FBM, a misaligned mesh strategy is proposed where the layout is divided in a set of non-overlapping rectangles as in Fig. 5.5b. In 3D, each rectangle corresponds to a cuboid. Each cuboid face is either a boundary region (i.e., the edge of the IC chip) or a side of adjacent cuboids. In order to apply the FDM scheme, a mesh point is defined inside each cuboid. Each substrate mesh node is then connected to the adjacent ones with the respective substrate parasitic elements.

The connection between meshing nodes of different doping material and level is done by means of diodes and homojunctions, while the connection between nodes of the same doping material is done with diffusion resistors. The geometrical size of each mesh cuboid is computed and defines the geometrical parameters of the component (area and length) of the parasitic devices.

The substrate meshing process consists of four steps starting with (1) the selection of the layout area of interest for meshing and (2) the creation of a list of layers. Each element of the list of layers is then meshed individually to build (3) the surface layout mesh. The two-dimensional meshing is then extended to the third dimension to build (4) the three-dimensional mesh.

5.3.2 Layout Area Selection

Accurate substrate parasitic coupling simulations impose meshing of the whole substrate. However, if only coupling effects between power devices and sensitive analog blocks are of primary concern, a selected portion of the layout can be defined to preserve the critical substrate coupling paths.

Coupling effects are responsible for disfunctioning of sensible analog blocks such as bandgap voltage reference circuit [7]. In this case, the extractions process concerns only the region of the layout including the bandgap reference and the aggressor circuits, i.e., the power transistor. Selection of the layout area to be extracted is done by drawing an appropriate layer, called SELECT, that covers the region of interest.

After the layout simplification process described in Sect. 5.2, the layout reverts to a set of geometrical shapes of p- and n-type layers. A shape recognition routine collects all the shapes included in the SELECT layer and creates two lists of layers according to their doping type obtained from Boolean layer operations:

- DN_All = DN AND SELECT
  - List of DN layers that will be extracted
- DP_All = DP AND SELECT
  - List of DP layers that will be extracted

5.3.3 Lists of Layout Layers for Meshing

The three-dimensional drift-diffusion model described in Chap. 3 is derived from a uniformly doped meshed volume. In order to extend the model to an integrated circuit containing many p- and n-type wells, each well must be processed and meshed independently. Then, every meshed region is connected with the other with the boundary elements, the EPFL diode, or the homojunction. This is why new lists of p- and n-type layers resulting from geometrical Boolean operations (AND, OR, NOT, XOR) applied to DN_All and DP_All lists are created.

Moreover, the purpose of each layer impacts how the new lists of layers are generated. Considering the cross section illustrated in Fig. 5.2, the DP layer is either drawn above the substrate to create deep substrate connections or above the DN to define the drift regions of LPDMOS or the body region of the complementary LNDMOS. The DP_All list is hence divided into two sub-lists: DP_IN_SUB is the list of DP layers formed within the substrate, while DP_IN_DN is the list of DP layers formed within the DN.

For instance, taking into account all the combinations of layers in an interdigitated NLDMOS, new lists of layers are generated (see Fig. 5.7) and summarized in the following steps:

- DN_WO_DP = DN_All XOR DP_All
  - List of DN layers not covered by DP
- DP_IN_DN = DP_All AND DN_All
  - List of DP layers covering DN
- DP_IN_SUB = DP_All XOR DP_IN_DN
  - List of DP layers in PSUB
- SUB_EX = (SELECT XOR DN_All) XOR DP_IN_SUB
  - List of PSUB layers

A merge operation is provided so as to join all the shapes within the lists created. In some technologies, the layout of the lateral DMOS transistor uses circular geometries to increase the switching speed and breakdown voltage [5]. Annular geometries and convex polygons are approximated by their bounding boxes for simplification. This ensures that each element is a rectilinear polygon with or without holes.

5.3.4 *Two-Dimensional Surface Meshing*

Meshing is first applied to the horizontal domain (x,y) of the layout. To illustrate the meshing methodology, a simple layout with many rectilinear polygons is considered. Figure 5.7a shows the simplified layout of an interdigitated LNDMOS, with the lists of layers including rectilinear polygons, with and without holes (Fig. 5.7b).

Meshing is done individually to each polygon included in the lists. This is done by applying a MESH2D function that approximates every rectilinear polygon (with or without holes) with a finite number of non-overlapping rectangles. To give a concrete example, the MESH2D function was applied to the DN_WO_DP polygon in Fig. 5.8a. The original polygon vertex is shown as black dots. To divide the polygon into rectangles, new vertex is derived during the meshing procedure. These are shown as white dots.

Those horizontal and vertical sides of the polygon, that enter in the polygon when they are stretched, are extended to derive the new vertex [8, 10]. The new vertex is placed at the intersection of the extended line with the vertical and horizontal boundary lines of the polygon. The center of each rectangle is the mesh node (Fig. 5.8b).

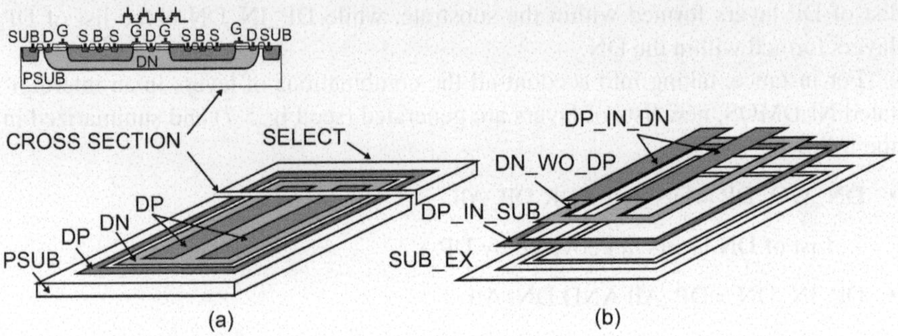

Fig. 5.7 (**a**) Simplified layout of an interdigitated LNDMOS transistor. (**b**) DP, DN, and PSUB layers are processed to build the lists of rectilinear polygons

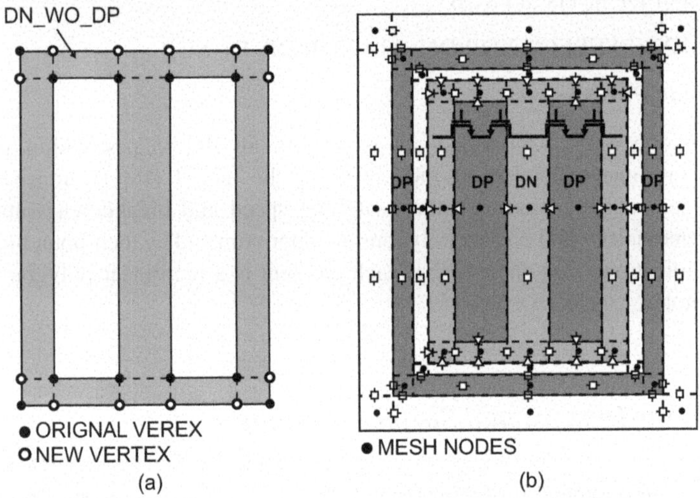

Fig. 5.8 (**a**) Application of MESH2D on DN_WO_DP polygon of Fig.5.7b. (**b**) By applying the MESH2D function to each polygon, the full layout mesh is obtained. The black dots represent the mesh nodes needed to build the substrate model

The pseudo code (see Algorithm 1) summarizes the two-dimensional layout meshing routine applied to each element of the lists of layers. By looping the MESH2D function over each list of polygons, the complete two-dimensional mesh of the layout is automatically generated. The meshing process is fast and ensures that the number of nodes is kept under control in layouts with nested polygons. Each rectangle generated by the meshing function can be further divided into smaller rectangles when more accuracy is needed.

Algorithm 1 Two-dimensional layout meshing routine

INPUT: Set of rectilinear polygons(P): P is a rectilinear polygon in the list
OUTPUT: List of Rectangles (R) and Mesh node (M): R is a rectangle in the list and M is a node in the list.

```
for all Polygon P_i in set P of rectilinear polygons do
    Divide the polygon in rectangles and fill the (R) list through MESH2D function
    for all rectangle R_j in R do
        Calculate the mesh point coordinates and fill the (M) list
    end for
end for
```

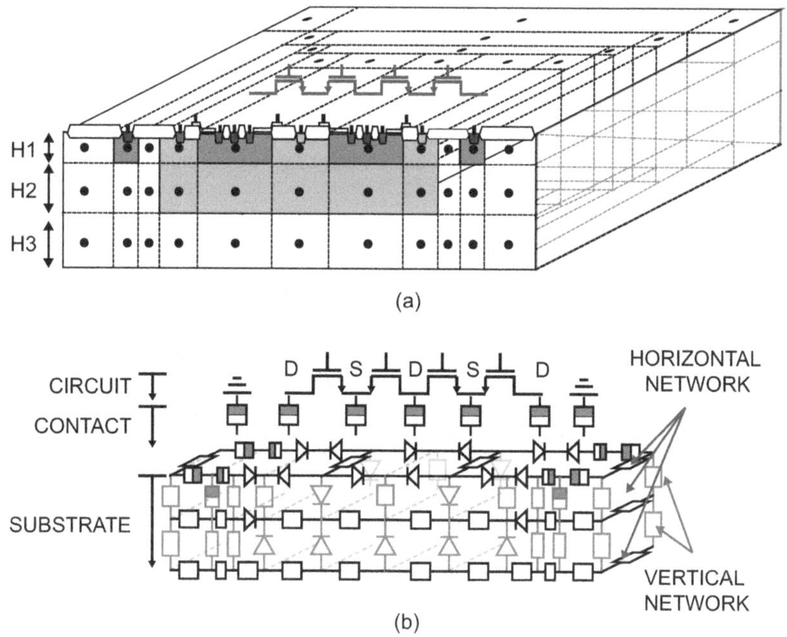

Fig. 5.9 3D Meshing of the layout (**a**). The substrate model is obtained by interconnecting both horizontal and vertical networks in order to include the surface circuits (**b**)

5.3.5 *Three-Dimensional Meshing*

The final three-dimensional meshing of the IC layout is built by interconnecting 2D mesh nodes in the normal direction. The initial two-dimensional meshing is replicated many times to build the 3D meshing according to the level of discretization desired in the vertical direction. In the example of Fig. 5.9a, the two-dimensional meshing was replicated three times to build the 3D meshing network.

Additional reduction of the number of nodes in 3D meshing is possible. Instead of simply replicating the 2D meshing in the z direction, the same approach used for 2D meshing can also be implemented.

5.4 Substrate Netlist Extraction for SPICE Simulation

The substrate parasitic model is constructed by interconnecting the resulting nodes to generate a grid-connected network (each node is connected to the neighboring nodes). A LVS checking software is used for the extraction and generation of the substrate network from layout data.

For this purpose, the technology extraction rule file is modified to automatically identify the substrate parasitic diodes, homojunctions, and diffusion resistors with their geometrical parameters and connectivity. This takes place during the LVS verification and results in a netlist that includes the circuit and substrate parasitics. The circuit netlist matches the original circuit schematic, while the substrate netlist contains the parasitic devices and connectivity based on the two-dimensional mesh grid. At this stage, netlist post-processing is used to build the three-dimensional schematic in order to complete the substrate extraction process. The substrate extraction is fast and relies on the efficiency of the layout verification tool.

5.4.1 3D Substrate Netlist

Vertical parasitic devices are extracted according to the "vertical" interactions between layers as shown in Fig. 5.2. In a given technology, in addition to the substrate, there are as many layers as wells of different junction depth.

In the example of Fig. 5.9a, the substrate cross section is characterized by three vertical layers having two wells of different depth (DP and DN). The final three-dimensional substrate network is formed by superposition of these three layers. The final 3D netlist is constructed by adapting the initial two-dimensional horizontal netlist and by taking into account the height of each layer (H1, H2, and H3) to calculate the geometrical parameters of the parasitic components. Subsequently, additional vertical netlists are generated to connect the horizontal netlists (Fig. 5.9b).

5.4.2 Schematic and Substrate Model Back Annotation

The interface between the circuit and parasitic substrate network is a fundamental aspect in co-simulation of parasitic signals with the circuit elements. The first step along the back-annotation process is to link the schematic circuit net names to the layout-extracted network. This is an important step, as the schematic net names are the ones used in the simulation of the circuit.

In the IC layout, the contact mask sets the connections of the circuit devices and corresponds to the substrate and to the p- and n-type wells. The superposition of a diffusion mask, a contact opening mask, and metal mask sets the connection between the metal and the underlying well. To back-annotate the substrate model to the schematic circuit, each circuit net name corresponding to a substrate contact is connected to the nearest mesh node in the grid through a parasitic component, most likely a homojunction. The contact area is defined by the diffusion mask area. Therefore, the electrical interactions between the circuit and substrate can be monitored during the simulations while maintaining the original schematic net names. The top vertical netlist of Fig. 5.9b makes the interconnection between the substrate network and the circuit. At this stage, the substrate extraction process is complete.

5.5 Implementation in Cadence Virtuoso Layout Editor

The extraction tool was developed to be fully compatible with EDA verification softwares (i.e., Cadence® Custom IC tool) and designed with an intuitive user interface GUI. The tool was integrated in the Virtuoso layout editor of Cadence and accessible via a drop-down menu.

The extensive set of geometrical functions available in Cadence® for LVS processing was exploited to efficiently extract the substrate model from a generic IC layout. In this way, parasitic interactions between circuit components in the substrate can be simulated during post-layout simulations with SPICE, preserving the back-annotation of substrate elements with the active circuit components.

In a pre-processing phase, once the configuration files (as described in Sect. 5.2 with technology geometrical and doping parameters) are loaded, the layout area within the drawing layer SELECT is meshed according to a minimal mesh size. As detailed in Sect. 5.4, the LVS is run to extract the 2D netlist which is the basis for building the complete 3D substrate model. During the post-processing phase, the substrate equivalent schematic is extracted.

5.5.1 Visualization of Results

In order to visualize SPICE simulation results, substrate node voltages and contact currents are saved at each simulation step. Substrate voltage distribution can be visually examined with a 2D color plot. The example of Fig. 5.10b shows substrate voltages for each vertical level of the meshed structure of Fig. 5.10a. This is useful to visualize substrate voltage profiles and to understand the de-biasing effects that could occur under the injection of a current from the parasitic vertical PNP transistor.

Similarly, the current distribution on the chip surface is determined by monitoring the current through each contact connecting the substrate with active devices. The color plot for the currents flowing in and out of the substrate contacts is shown on top of Fig. 5.10b. Critical current paths due to the activation of lateral NPNs can then be identified during post-layout simulations.

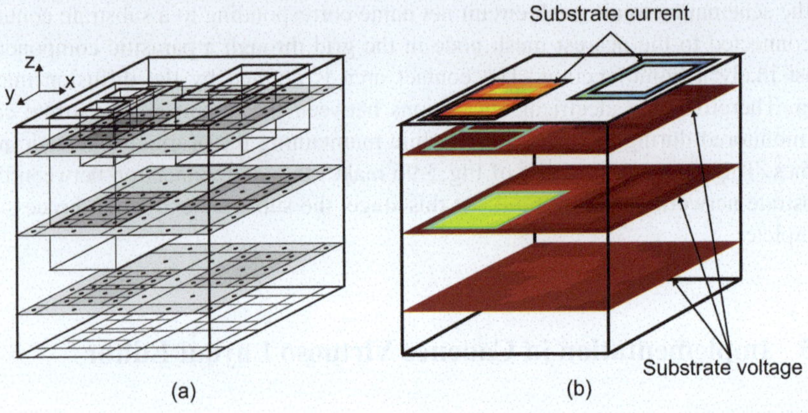

Fig. 5.10 3D meshing (**a**) with substrate voltage and current distribution issued from SPICE simulations (**b**)

5.6 Substrate Current and Grid Resolution

In this section, several simulations are compared to estimate the impact of the grid resolution on the simulated substrate current. For this numerical study, the grid cell size varies from a 100 μm width (basic) as shown in Fig. 5.11a down to 7 μm as shown in Fig. 5.11b. Moreover, the number of vertical layers ranges from one to four levels. The size of the benchmark layout size is 700 μm × 400 μm. The monitoring parameter is the collected current measured at C1 and C2 for an injected substrate current of 10 mA at the emitter E. Simulation results for C1 and C2 currents are shown, respectively, in Fig. 5.11c and d over the extracted number of nodes. The experiment consists of 20 simulations.

Figure 5.11c suggests that for a given number of vertical substrate layers, there is a slight increase of the C1 current as the grid size decreases. Surprisingly, C1 current is nearly constant when only one vertical layer is considered. There is, however, a significant increase of C1 current when the number of vertical layers increases. Moreover, this analysis suggests that the lateral substrate current is more sensitive to the vertical grid resolution than the horizontal one. This result is expected since the substrate current can flow deeply into the substrate.

The current flowing through the collector C2 is almost constant, regardless of the number of nodes (Fig. 5.11d). In this example, C1 is closer to the emitter and collects most of the injected current and also acts as a barrier with respect to C2. The current C2 increases as well as with the number of vertical layers, and the maximum variation occurs when the number of substrate layers changes from one to two. In Fig. 5.11c the simulation time is also reported. The computation time for the model with the higher number of nodes increases significantly (several order of magnitude) with respect to the simpler model. Maximizing the accuracy and minimizing the simulation time involves a trade-off. Importantly, these results

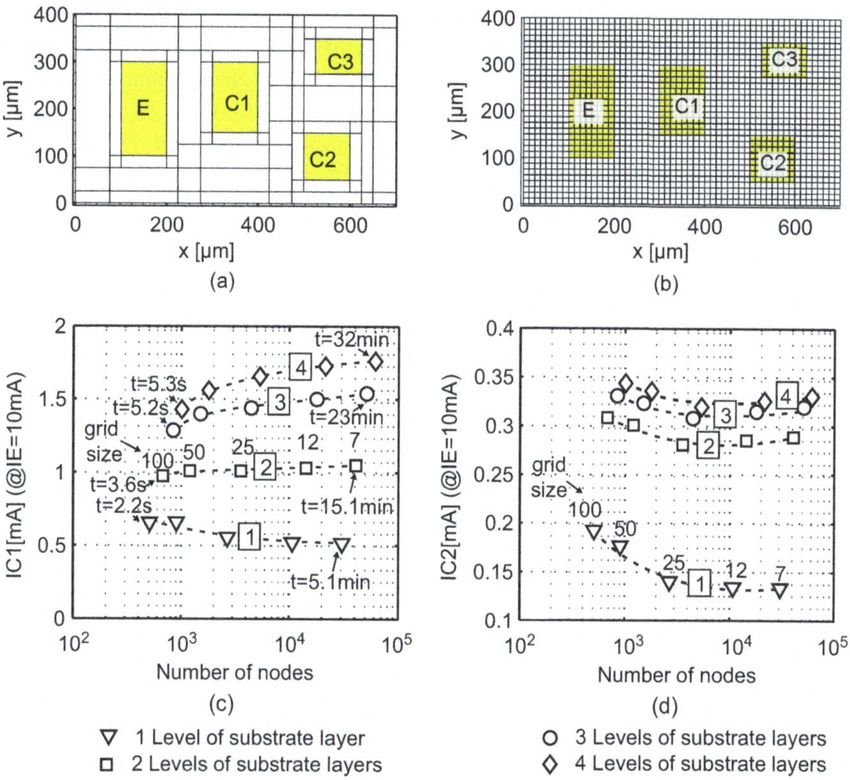

Fig. 5.11 Layout with 100 μm grid (**a**) and 7 μm grid (**b**). Simulation results of IC1 (**c**) and IC2 (**d**) currents versus number of extracted node

indicate that increasing the vertical grid resolution, rather than the horizontal one, leads to an acceptable trade-off between accuracy and simulation time.

5.7 Conclusion

This chapter presents a tool to extract the three-dimensional EPFL substrate model from a given circuit layout. The tool is a major improvement in estimating coupling signals in high-voltage ICs due to parasitic BJT transistors during the usual mixed signal flow. The tool enables simulation of substrate couplings for any chip configuration and technology. The layout mesh generation is an important step toward the generation of a substrate network model given that the number of nodes will considerably impact the network complexity and increase the simulation time. The novel layout mesh generation algorithm proposed divides the circuit geometry into rectangular cells to ensure an optimal partitioning of the layout while keeping the propagation of the mesh nodes under control, especially in layouts with nested structures. In practical situations, only a portion of the IC layout substrate can be selected and extracted by means of a selection layer. This ensures that the number of nodes of the substrate model is under control and prevents convergence issues and excessive simulation times. With an easy-to-use interface and visual representation of the simulated data, this tool can be used in place of a device simulator to predict substrate couplings in high-voltage circuits.

References

1. H. Ballan, M. Declercq, M. Declercq, in *High Voltage Devices and Circuits in Standard CMOS Technologies* (Springer, New York, 1999)
2. P. Buccella, C. Stefanucci, H. Zou, Y. Moursy, R. Iskander, J.-M. Sallese, M. Kayal, Methodology for 3-D substrate network extraction for spice simulation of parasitic currents in smart power ICs. IEEE Trans. Comput.-Aided Des. Integr. Circ. Syst. **9**, 1489–1502 (2016)
3. F.L. Conte, J.-M. Sallese, M. Pastre, F. Krummenacher, M. Kayal, Global modeling strategy of parasitic coupled currents induced by minority-carrier propagation in semiconductor substrates. IEEE Trans. Electron Devices **57**(1), 263–272 (2010)
4. A.F. Franz, G.A. Franz, S. Selberherr, C. Ringhofer, P. Markowich, Finite boxes - a generalization of the finite-difference method suitable for semiconductor device simulation. IEEE Trans. Electron Devices **30**(9), 1070–1082 (1983)
5. A. Hastings, *The Art of Analog Layout* (Prentice Hall, Upper Saddle River, 2005)
6. H.-P. Hong, J.-C. Wu, A reverse-voltage protection circuit for MOSFET power switches. IEEE J. Solid State Circuits **36**(1), 152–155 (2001)
7. W. Horn, H. Zitta, A robust smart power bandgap reference circuit for use in an automotive environment. IEEE J. Solid State Circuits **37**(7), 949–952 (2002)
8. P.-Y. Hsiao, C.-Y. Lin, P.-W. Shew, Optimal tile partition for space region of integrated circuits geometry. IEEE Proc. Comput. Digit. Tech. **140**(3), 145–153 (1993)
9. S. Li, Y. Fu, Smart power technology and power semiconductor devices, in *Devices and Optoelectronics 3D TCAD Simulation for Semiconductor Processes* (Springer, New York, 2012), pp. 187–236
10. V.H. Nguyen, Optimum partitioning of rectilinear layouts. IEE Proc. Comput. Digit. Tech. **143**(6), 440–442 (1996)
11. C. Stefanucci, P. Buccella, M. Kayal, J.M. Sallese, Impact of enhanced contact doping on minority carriers diffusion currents, in *10th Conference on Ph.D. Research in Microelectronics and Electronics (PRIME), 2014* (IEEE, Grenoble, 2014), pp. 1–4

Chapter 6
Parasitic Bipolar Transistors in Benchmark Structures

6.1 Technology Calibration Methodology

Model parameters should be calibrated with respect to the selected technology process. However, the intrinsic distributed and three-dimensional feature of the EPFL substrate model compared to semiconductor device compact models asks for specific parameter extraction methodologies. For instance, extraction routines for a set of devices with different dimensions (e.g., in a transistor with different width W and length L) are not applicable for the substrate model.

In the proposed substrate modeling, all geometrical aspects are taken into account by the meshing algorithm; thus, specific test structures have to be selected for calibration purposes where only the technology parameters (mainly dopings and lifetimes) should be extracted.

The technology calibration methodology for the distributed model of the substrate is proposed in Fig. 6.1 and consists of three steps:

- **Technology information:** this is required to have process information from the foundry regarding the cross section. In particular the mask layers, the junction depths, the epi-layer thickness, or the deep trench length must be known. From SPICE models available in the PDK, diodes, BJTs, and resistor parameters can be used to estimate some technology features. This is how the average doping is estimated and used as a starting value for the calibration procedure.
- **Calibration structure layout and meshing:** selected calibration devices are laid out. They consist of diodes, bipolar transistors, or parasitic multi-collector structures whose corresponding layout is meshed to extract the equivalent substrate model. Since the meshing algorithm can affect calibration, a fine mesh must be adopted to minimize possible errors. The drawback is an increase in simulation and calibration time.

P. Buccella et al., *Parasitic Substrate Coupling in High Voltage Integrated Circuits*, Analog Circuits and Signal Processing, https://doi.org/10.1007/978-3-319-74382-0_6

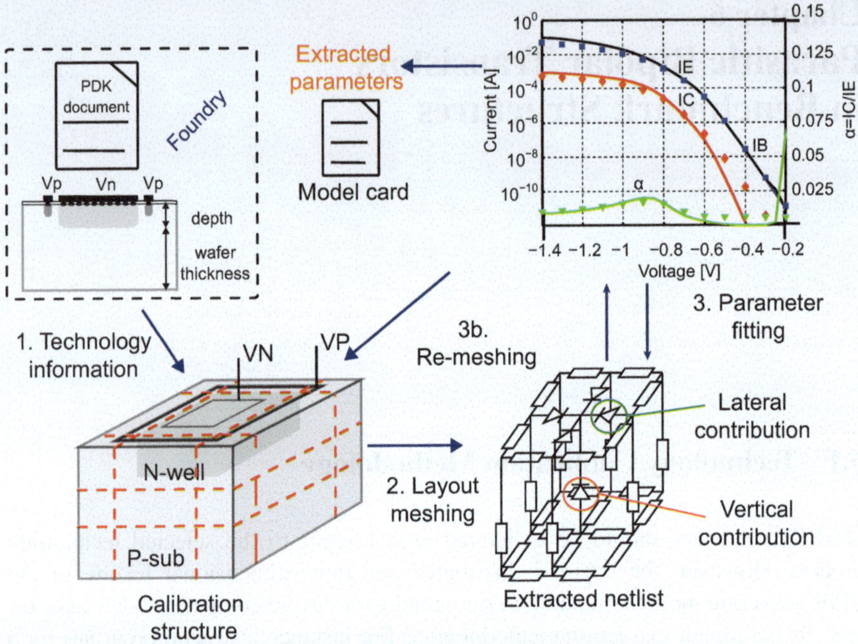

Fig. 6.1 Technology calibration methodology flow for the EPFL substrate model

- **Parameter extraction and re-mesh:** the calibration devices are measured, and parameter sweeps are performed to match the equivalent substrate model simulations with measurements. Compared to standard extraction methodologies, this task is performed with a netlist including all EPFL components at the same time. Appropriate ranges for the parameters must be fixed, and several interactions must be run for the final calibration by extracting few parameters at a time. Assessment with respect to the mesh should also be performed.

At the end of the calibration process, a model card file stores all the technology-dependent parameters to be used in substrate coupling simulations. In this work an optimal calibration routine was not developed since it requires a flexible and automatic substrate extraction tool as well as specific studies on the most appropriate test structures. Nonetheless, an example of the calibration methodology is reported for a JI HV-CMOS technology without backside contact by using common devices like diodes and simple parasitic BJTs.

In the following graphs, continuous lines refer to circuit simulations and dots to measurements. All the parameter optimization routines are implemented in Keysight IC-CAP software by using the Levenberg-Marquardt algorithm. Note that for a distributed substrate model, the input netlist was modified in order to be compatible with the IC-CAP input format.

6.1.1 Diode Measurements

As simple device structures, diodes are the best candidates for the parameter extraction of substrate devices. Diodes are usually foundry test structures or parasitic devices in all MOS transistors. Since many PN combinations are available in a real technology, only the most relevant PN junction for parasitic lateral couplings will be considered as an example, i.e., the deep n-well (DN) to p-substrate diode.

For calibration purposes, the diodes must be characterized in reverse and forward bias conditions. To measure the reverse current, large structures are needed (increasing however the simulation time). Moreover, two calibration structures as the ones presented in Fig. 6.2 are considered. The first one is a $800 \times 800\,\mu m^2$ n-well surrounded by a P+ contact ring, while the second has multiple wells $800 \times 20\,\mu m^2$ connected in parallel. The first diode is called the *area device*, while the second one emulating the parasitic structure in a fingered transistor is called the *perimeter device* since the resulting n-well perimeter is much larger. The same devices are used for SPICE diode model calibration where there have different parameters for the vertical area and the lateral perimeter contributions. In the proposed substrate

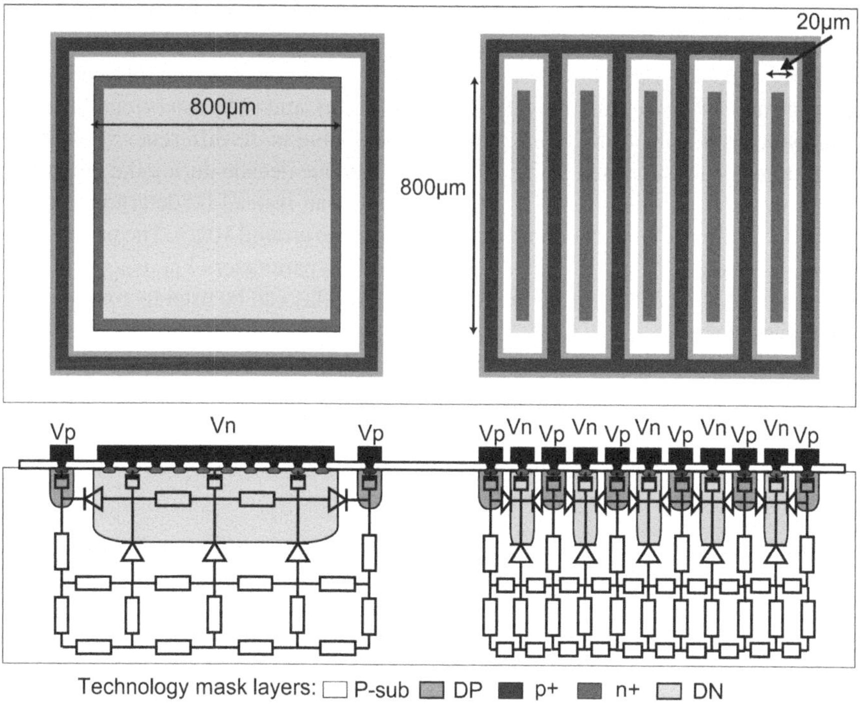

Fig. 6.2 Top layout view and cross section of DN-PSUB diode calibration structures for area device (left) and perimeter device (right). The EPFL substrate model is sketched for the two devices in the cross section (drawings not to scale)

network, each contribution is automatically identified since different devices are used for the vertical and lateral diodes (see Fig. 6.1).

Concerning the calibration procedure, the DN junction depth and the substrate thickness are data provided by the foundry. In the selected technology, a uniformly doped p-substrate is used. After the layout extraction, the substrate was meshed in cuboids with maximum size of 50 μm along the three dimensions. The parasitic network includes diodes at PN junctions, resistors inside the p-substrate and the DN well, and homojunctions at PP+ and NN+ contacts. Note that for the area device, the equivalent netlist results in 11,120 components, while for the perimeter device, it is composed of 2700 lumped devices. As reported in the cross section of Fig. 6.2, the fingering process increases the number of parasitic components as more resolution is required.

The parameters to extract are the average doping concentrations for the substrate (N_a), for the P+ and N+ contacts (N_{a+}, N_{d+}), and for the DN well (N_d), as well as the electron-hole pair lifetime τ_0. Due to the process and the doping profile, the parameters are different between vertical and lateral lumped devices. However, this is true for all phenomena involving the depletion region, i.e., for leakage currents and AC coupling. For the reverse current contribution, two different recombination lifetimes for the vertical and lateral diodes (τ_{rv}, τ_{rl}) must be determined. Also the AC parameters (V_{bi}, x_0, m) are different for the lateral and vertical dimensions.

Since there are many parameters and only two calibration devices, a procedure must be selected. First of all, from the technology information inspection of diffusion and substrate resistivity, Eqs. (A.5a), (A.5b), and (3.9) can be combined to roughly determine the order of magnitude of the doping in the different wells. These initial dopings are supposed to change by less than one decade during the optimization process. The initial guess for AC parameters can instead be determined from the SPICE compact model keeping the lifetime values around 10 μs. The parameters V_{bi}, x_0, m are indeed exactly equivalent to the SPICE parameters $V_{\mathrm{BI}}, C_{J0}, M$ for the junction capacitance. Therefore the following formulas can be used to compute the initial guess values:

$$V_{bi} = V_{\mathrm{BI}} \quad m = M \quad x_0 = \frac{\varepsilon_s}{C_{J0} V_{\mathrm{BI}}^M} \tag{6.1}$$

All the calibration steps hereafter use the measurements on the two devices (area and perimeter) at the same time to fasten convergence:

- **Step 1:** AC simulation of the diodes in reverse bias is run to optimize the CV characteristics of the model with measurements. The final result is reported in Fig. 6.3.
- **Step 2:** At this point the DC parameters are extracted. In reverse bias condition, the SNS contribution is dominant for silicon, meaning that the reverse bias current can be effectively used to extract the recombination lifetimes τ_{rv}, τ_{rl} for the lateral and vertical devices.

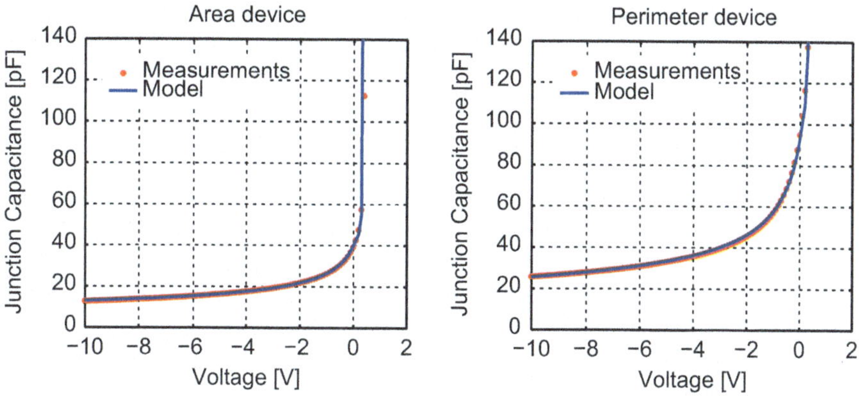

Fig. 6.3 CV measurements (dots) versus simulated characteristics (lines) for the diodes DN-PSUB of Fig. 6.2

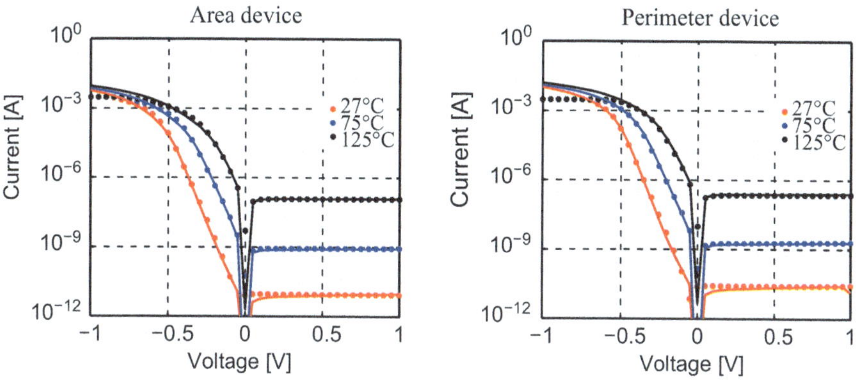

Fig. 6.4 *IV* measurements (dots) versus simulated characteristics (lines) for the diodes DN-PSUB of Fig. 6.2 at three different temperatures

- **Step 3:** The doping parameters N_a, N_d can be extracted in forward bias condition. It is important to mention that at very high bias, the current will be limited by the parasitic series resistance of metal interconnections or by the measurement setup. It is therefore compulsory to correctly estimate this resistance and include it during the simulations. In our case the area device has a parasitic resistance of 40 Ω, and the perimeter device of 30 Ω is limiting the current in high injection.

After optimization the results shown in Fig. 6.4 are obtained. Note that for safety issues, the current compliance of 3 mA was set during measurements resulting in clamped values for high forward bias regime. However, at the end of the calibration procedure, N_{a+}, N_{d+}, τ_m must still be determined. As a matter of fact, no high sensitivity is obtained for these parameters, and often the extracted values are not optimized. This means that additional structures are required.

A final check with temperature sweep shows how the implemented temperature dependence is well modeled by considering only the lifetime temperature coefficient of about $\alpha = 1.5$.

6.1.2 Parameter Tuning on Simple Structures

Specific test structures for lateral parasitic configurations can be used for parameter extraction. To this purpose simple aligned DN wells are designed that represent a multi-collector NPN BJT. In the designed chip, the area of each well is about 14,000 μm^2 in order to scale the reverse current to measurable values (see Fig. 6.5).

In bipolar transistors not only the absolute value of currents as in the Gummel plot is used for parameter optimization but also the transport factor $\alpha = I_C/I_E$. This parameter is very important to quantify substrate couplings between different wells. However, being related to a current ratio, matching with measurements is challenging, and no substrate model reported in the literature so far demonstrated a good α-matching. The model simulations after α-calibration are shown in Fig. 6.5 for the two NPN combinations between N2(emitter)-N1(collector) and N2(emitter)-N3(collector). The model is able to track both the small difference of α due to the slightly different distance and the α variation with respect to different collector voltages (in the example 0 and 10 V). Note that the high injection variation of α is also strongly dependent on the parasitic metal contact resistance that is always included in the simulations.

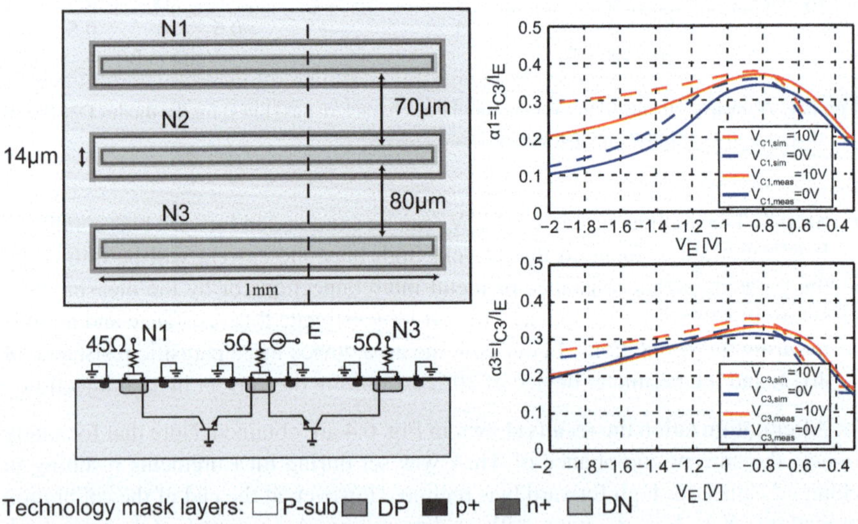

Fig. 6.5 (Left) Top view and cross section of parasitic NPN BJT's calibration structures, not to scale. (Right) Calibrated α parameters at different collector voltages for two parasitic BJTs

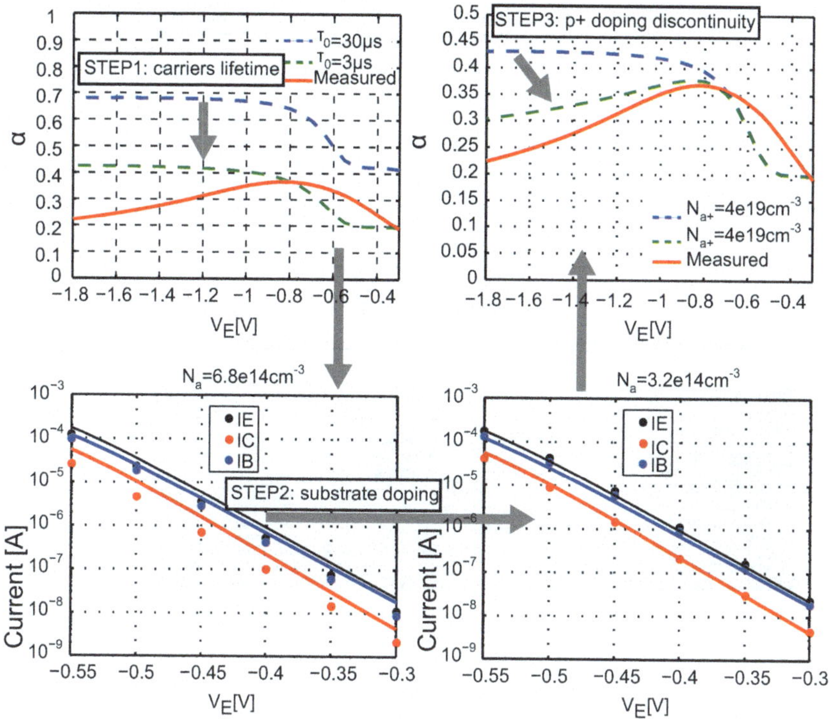

Fig. 6.6 Three-step selective parameter optimization for lateral parasitic NPN BJT

For parameter optimization, the substrate is the most sensitive element as expected for lateral parasitic couplings. The substrate average doping N_a, the electron lifetime τ_0, and the P+ contact doping N_{a+} are consequently the most sensitive parameters on the results. Instead of optimizing the three altogether, selective optimization can be done. As reported in Fig. 6.6, the aforementioned parameters have different effects. First, the lifetime can be adjusted to scale the coupling α between the emitted and the collected currents. Then the substrate doping is optimized to match the absolute value of the injected current in the Gummel plot (especially in the low injection, diffusion-driven regime). Finally the P+ interface has a role for shaping the α in the high injection regime.

6.2 Model Verification on Simple Test Structures

The predictive behavior of the substrate model was verified on a specific test chip designed and fabricated in 0.35 μm HV-CMOS process to investigate different guard ring layout options. The chip layout is shown in Fig. 6.7a. It is composed

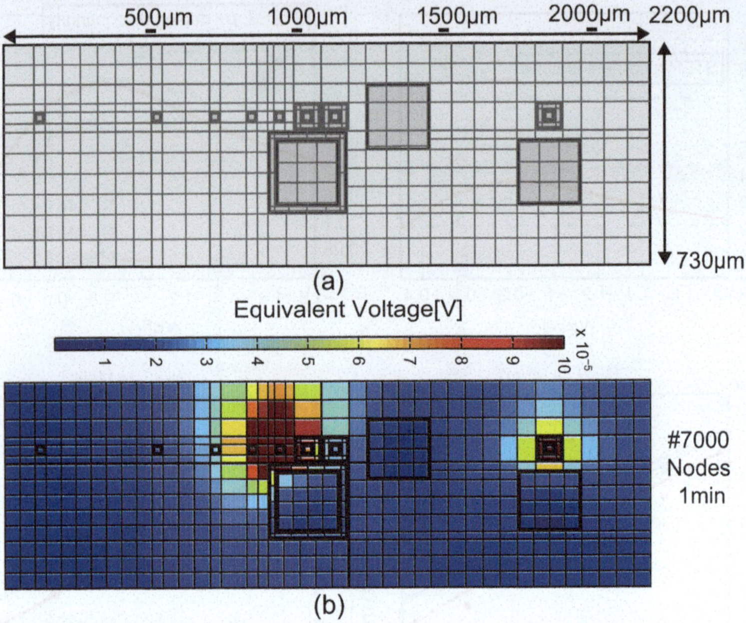

Fig. 6.7 (**a**) Test chip layout with corresponding surface meshing. (**b**) A simulation with two emitting nodes is also shown where the color map is proportional to the electron concentration into the substrate

of different arrangements of n-wells and p-rings of various dimensions and orientations. Some measurements and simulations were selected to show the activation of various parasitic lateral multi-collector NPN transistors. The corresponding surface nonuniform mesh used is also reported in Fig. 6.7a. Regarding the vertical mesh, it was done with five binary scaled subdivisions. In fact since the backside contact is not present, the parasitic current propagation is mainly a surface effect. The substrate layer near the top contacts is made thin for better accuracy, while deeper layers are coarse since their contribution is smaller. However, the optimal number of vertical layers depends on the simulated area and on the technology. Since a large layout area is simulated (about $2\,mm^2$), all the substrate thickness is considered.

The resulting number of electrical nodes in the netlist and average simulation time for a sweep of 10 DC points is reported for completeness in Fig. 6.7b. The size of the netlist increases with the area, while the simulation times still remain in a good range for practical use. The measurements are performed on-wafer to reduce parasitics. However the effects of the metal trace resistance were taken into account in the following simulations.

Note that with the proposed model, the electron concentration into the substrate can be easily visualized with a color map plot as in Fig. 6.7b. This captures node voltage and current variations in the substrate when two n-wells are injecting electrons, but where one is properly shielded, while the other has no protection guard rings.

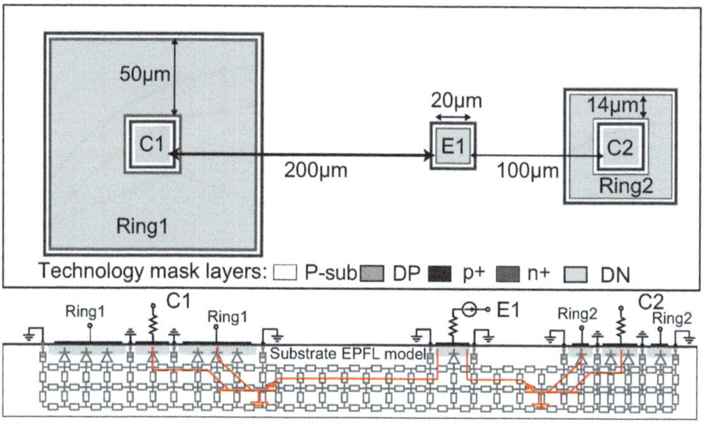

Fig. 6.8 Simple layout (top) and corresponding cross section with the EPFL substrate model (bottom) of a multi-collector parasitic NPN BJT to simulate the current propagation

6.2.1 *Effect of Distance and Guard Rings*

The simplest way to reduce the coupling from the substrate current is to increase the distance between aggressor and victim n-wells [2]. However, to reduce silicon area, n-type guard rings placed around victim n-wells are usually used as protection structures for minority carriers. To show the combination of distance and guard ring dimensions, the test structure shown in Fig. 6.8 is used. The emitter E1 injects a substrate current toward C1 placed 200 μm away with a shielding N-guard ring of 50 μm width or toward C2 with the same area but placed closer and with a smaller N protection ring (14 μm wide). Measurements and simulations were done by injecting a current into the substrate from the $20 \times 20\,\mu\text{m}^2$ emitting n-well E1 and by measuring the collected current at each collector biased at a constant DC voltage. Results are shown in Fig. 6.9. The model can track the magnitude of substrate couplings through the NPN transistor gain $\alpha = I_C/I_E$ from low to high current injection. As expected, no significant difference is reported for the α in each case since most of the current is collected by the protecting rings. The model can then be used to optimize the area.

6.2.2 *Guard Ring Placement*

In real circuit it is difficult to derive general rules that link the collector area with the required n-ring width. For this reason, the substrate model can be used to study the best combination of distance and guard ring dimension to target a certain robustness in a given layout. In the previous example, the protecting guard rings were placed only around the collecting n-wells. In the benchmark structure reported in Fig. 6.10,

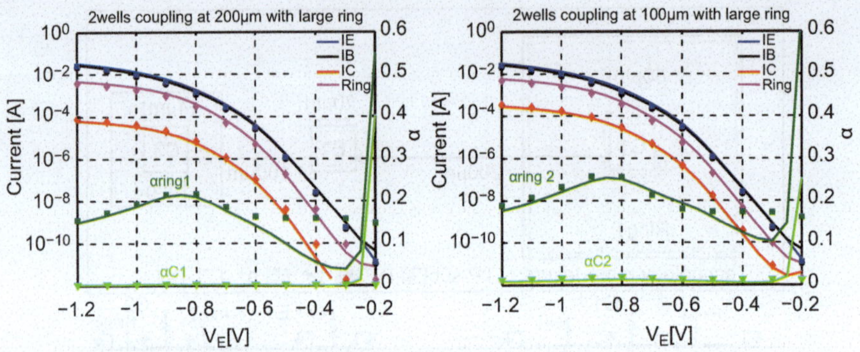

Fig. 6.9 Gummel plot and multi-collector α values of the lateral NPNs shown in the test structure of Fig. 6.8. Points correspond to model simulation and continuous line to measurements

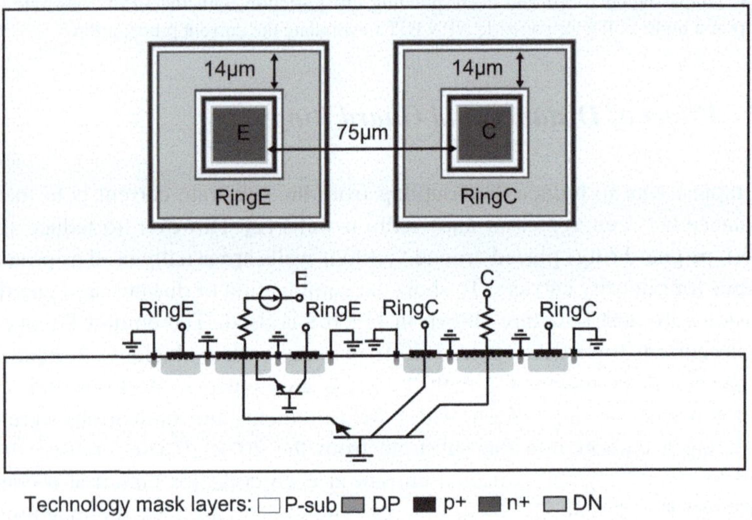

Fig. 6.10 Simulated aggressor-victim geometry with two n-well guard rings for minority carriers

where both emitting and collecting n-wells are surrounded by N-guard rings, the optimal placement of the rings is investigated.

The substrate coupling results are shown in Fig. 6.11. When the two rings are not biased, the relative couplings between the two n-wells placed 75 µm apart are around $\alpha = 0.14$ (the corresponding Gummel plot is not reported). Connecting the collector ring, the coupling is decreased to $\alpha = 0.014$ since the ring efficiency is around $\alpha_{\text{ringC}} \simeq 0.3$ (Fig. 6.11a). If instead the emitter ring is connected, the coupling is decreased to $\alpha = 0.01$, but in this case the ring efficiency increases up to $\alpha_{\text{ringE}} \simeq 0.8$ (Fig. 6.11c). This means that placing guard rings around emitter regions is much more effective than placing them at the collecting regions. Connecting both

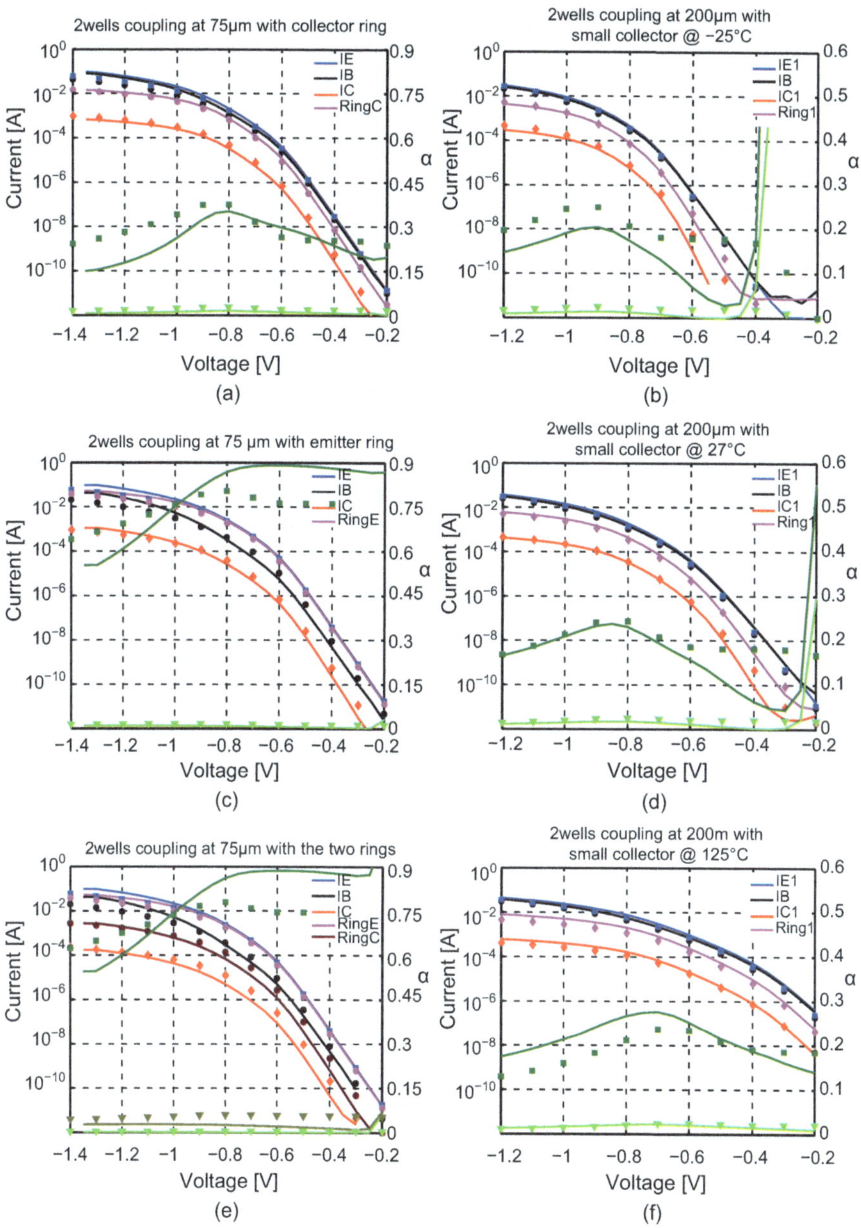

Fig. 6.11 (**a**), (**c**), (**e**) Gummel plot and α variation of parasitic multi-collector NPN BJT with different guard ring connections and (**b**), (**d**), (**f**) at different temperatures. Points correspond to the model simulation and continuous line to measurements

rings at the same time, the coupling between the two wells is decreased to safe values of $\alpha_{\text{ringC}} \simeq 0.005$, but the emitter ring efficiency remains of 80%, while the collector ring collects only 1% of the current (Fig. 6.11e). The same behavior has been demonstrated for much larger distances of about 800 μm.

In conclusion, when an aggressor source is detected, enclosing it with a shield should be investigated first before addressing additional configurations.

6.2.3 *Effect of Temperature*

The model can track temperature effects as well. The test structure is similar to the one in Fig. 6.10, where the emitter E1 ($20 \times 20\,\mu\text{m}^2$) is placed 200 μm away from a larger collector C1 ($100 \times 100\,\mu\ \text{m}^2$) enclosed by a 50 μ m wide n-type guard ring (Ring1). In this case the emitting well injects current that is collected by the shielded collector *C*1 at 200 μm. Simulated at room temperature (Fig. 6.11d), confirm again the shielding effect of the n-ring around the collector victims. Increasing the temperature at 125 °C, the injected current increases as expected (Fig. 6.11f). This increase is mainly visible in the low injection region driven by diffusion, while for the high injection region driven by drift, the series resistance is the main limiting factor. The opposite trend, i.e., a current decrease, is seen at −25 °C as predicted by the model (Fig. 6.11b).

6.2.4 *Lateral NPN: Double Emitter*

When multiple power transistors are integrated in the same IC (i.e., for multichannel driving applications), several injectors may be activated at the same time. To some extent, the current collected by the victims is the cumulative effect of the injected currents [3]. However, the interaction between the injectors changes with their bias voltage with respect to the substrate. Indeed, a NPN transistor can operate in different modes and can conduct current both in forward-active and reverse-active modes, depending on the emitter-base and collector-base junction bias.

To reproduce this situation, injection of a substrate current is done by forward biasing each substrate junctions *E*1 and *E*2 simultaneously. The respective simulation setup is shown in Fig. 6.12.

Terminals *E*1 and *E*2 are biased negatively with respect to the substrate with a parametric voltage sweep *VE*1 and *VE*2 in the range of 0 to −2 V. Measurement and simulation results are shown in Fig. 6.13a, b, where the current through *E*2 is plotted versus *VE*1 and *VE*2. The *E*2 current flow can change in direction: the current is positive when it enters terminal *E*2, but it becomes negative when it leaves the terminal. This means that *E*2 behaves as an emitter or collector, depending on *E*1 and *E*2 relative emitter biasing conditions with respect to the substrate.

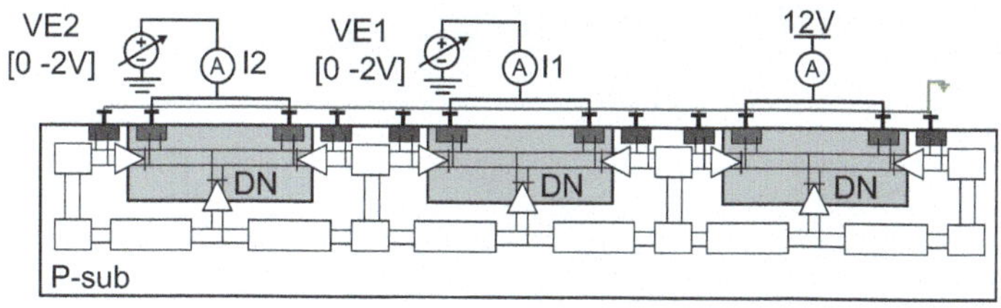

Fig. 6.12 Setup for measurement and simulation of injector effects. The equivalent substrate model for SPICE simulations is also shown

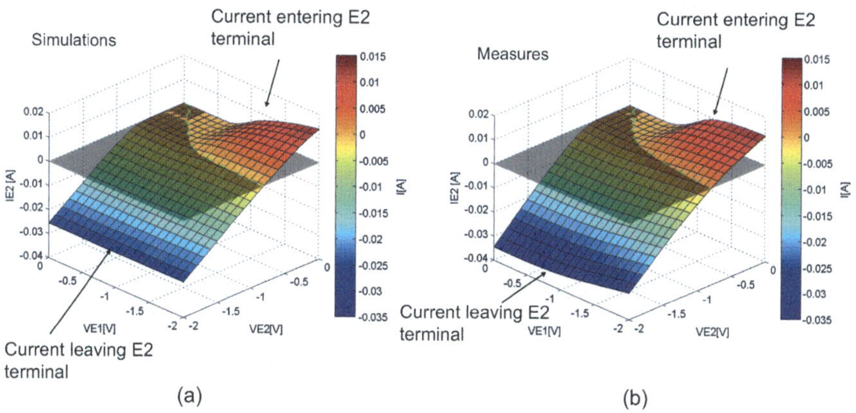

Fig. 6.13 (**a**) Simulation and (**b**) measurement results of *E*2 current in the double-emitter situation

Plots in Fig. 6.13 illustrate the smooth transition from forward-active to reverse-active conduction mode on the lateral NPN, which is predicted according by the substrate SPICE model.

6.3 Model Verification in Advanced Benchmark Structures

A test chip including an H-Bridge representing a real IC application was designed and fabricated in a 0.35 μm HV-CMOS technology. The chip photography is shown in Fig. 6.14; the die is 3.7 mm × 2.1 mm.

The H-Bridge includes four DMOS power transistors designed to drive 0.5 A with a power supply VDDH = 12 V. More specifically, the H-Bridge comprises two n-type LS NDMOSs (N1 and N2) and two lateral HS PDMOSs (P1 and P2). The four transistors are placed in the chip layout in the order P1, N1, N2, and P2 as is also shown in Fig. 6.14 covering an area of approximately 2.4 mm^2.

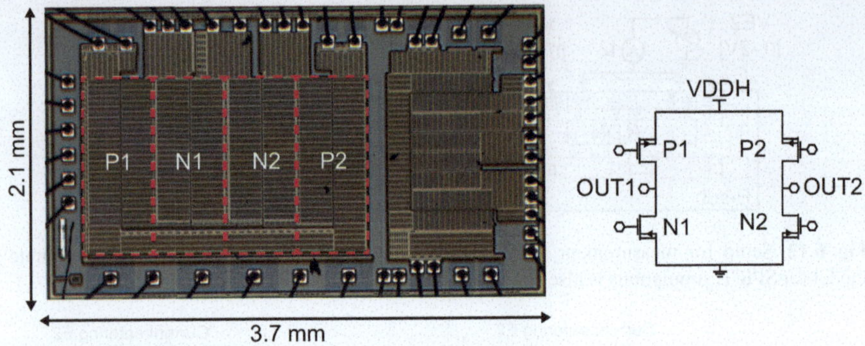

Fig. 6.14 Chip photograph of the test chip realized in 0.35 μm HV-CMOS; the die measures 3.7 mm × 2.1 mm

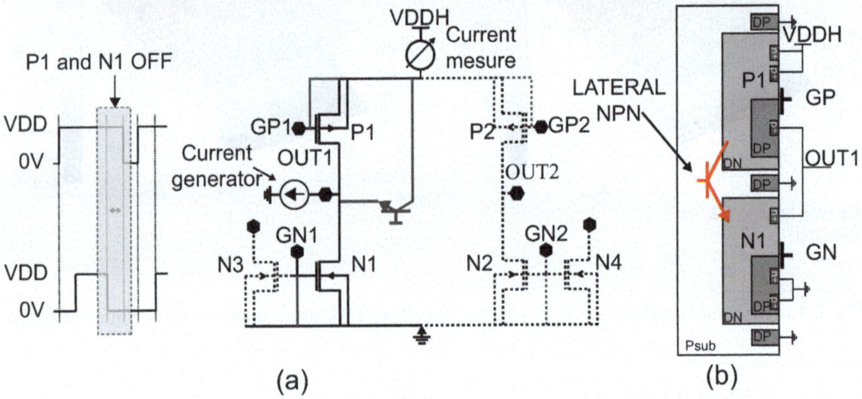

Fig. 6.15 Schematic of the Half-Bridge with complementary N- and P-channel power transistors (**a**). Power driver cross section showing the substrate lateral NPN parasitic device (**b**)

6.3.1 Half-Bridge Driver

In this section, the substrate coupling currents caused by the activation of the lateral NPN are presented through the extraction and simulations of substrate parasitics from the designed circuit layout. For the substrate parasitic model extraction, only two among the four H-Bridge transistors are considered. The two transistors are actually a Half-Bridge power driver. The equivalent schematic is illustrated in Fig. 6.15a and comprises the two complementary power switches P1 and N1 in a push-pull configuration. The Half-Bridge is capable of sourcing and sinking up to 0.5 A with a power supply of 12 V.

The Half-Bridge is typically used in applications designed to switch an inductive load. During switching, special care must be taken to control the inductor current [8]. Indeed, when the driver is off (P1 and N1 are simultaneously turned off), the inductive current drives the output node (OUT1) below the ground voltage, activating the substrate parasitic device that includes the lateral bipolar transistor (Fig. 6.15b).

The activation of such a device can affect the functionality of analog and digital blocks and even destroy the chip if a latch-up is triggered [4]. When the Half-Bridge output node (OUT1) goes below ground, it forward biases the substrate PN isolation junctions, and uncontrolled parasitic currents are injected into the substrate. These parasitic currents spread nonuniformly in all the directions throughout the substrate until they reach a low-voltage node of the circuit.

The power driver layout consists of the two high-voltage P- and NDMOS (P1 and N1) devices with an interdigitated source and drain (Fig. 6.16). The general layout structure includes the PDMOS transistor (P1) and the NDMOS transistor (N1) built of three sections and six sections, respectively. The sections of P1 are arranged in three DN wells spaced regularly according to the technology design rules. The source regions (123 fingers) are electrically connected to the P1 bulk and to the VDDH pad. Similarly, the multiple P1 gate regions are electrically connected to the GP output pad. In the same way, the drain regions (120 fingers) are electrically connected to the OUT pad.

The sections of N1 are also arranged in 6 DN wells with the source and bulk (108 fingers) connected to the GND pad, the gate to the GN pad, and the drain (120 fingers) to the OUT1 pad. A multitude of substrate contacts provide an equipotential biasing of the substrate to the ground. The Half-Bridge driver layout occupies an area of 1.17 mm^2.

In order to simulate substrate couplings between the driver output node and the power supply, the substrate area hosting transistors P1 and N1 was meshed and extracted (see Fig. 6.16). However, in order to minimize the number of nodes and hence the number of extracted devices, a prior circuit analysis is done, resulting in two layout meshing iterations named S1 and S2 based on two different layout reduction strategies. In both extractions, the source to body diodes of P1 and N1 transistors are ignored since they are short circuited by the metal connection.

- **S1 meshing strategy**. The low-side power NDOMS transistor parasitic junctions between the DN-PSUB layers and DN-DP layer are forward biased when the output goes below the ground level (see Fig. 6.17a). On the other side, the high-side PMOS PSUB-DN well junction is reverse biased and acts as the parasitic collector of the lateral NPN transistor (DN-PSUB-DN). The layout including the DN, DP, and PSUB layers is meshed and extracted.
- **S2 meshing strategy**. The goal of the S2 meshing strategy is to reduce the number of parasitic components. This strategy relies on the assumption that the vertical DP/DN/PSUB PNP is never activated during the simulation setup. The vertical PNP is indeed always in the cutoff region during the simulation, with both junctions (DP-DN and PSUB-DN) reverse biased. As a result, the DP-DN diode of the high-side power transistor is omitted from the extraction procedure (see Fig. 6.17b). Thus, in-line with this meshing strategy, only DP-DN diodes are extracted, and all DP-DN diodes of the power transistor P1 are omitted.

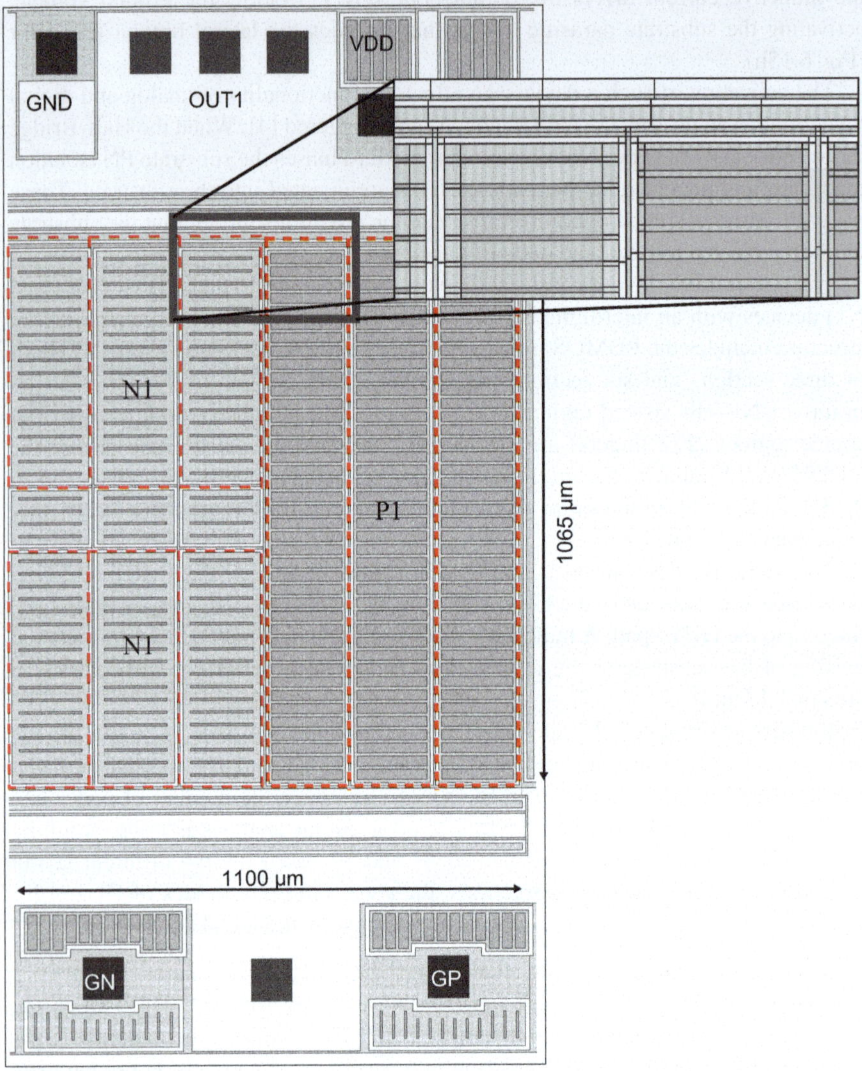

Fig. 6.16 Layout view of the Half-Bridge driver structure with a zoom of the mesh applied to N1 and P1 layout

Then two substrate networks are extracted. The extracted netlist number of nodes and elements with extraction times are reported in Table 6.1.

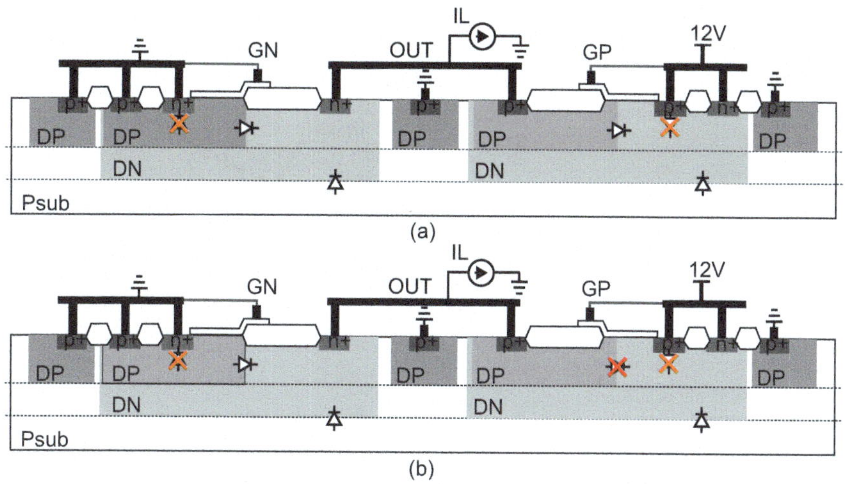

Fig. 6.17 Power driver cross section representing the S1 meshing strategy (**a**) and the S2 meshing strategy (**b**). Source to body diodes are not extracted in either the S1 or S2. Furthermore, in S2 the DP-DN diode of the high-side power transistor is not extracted

Table 6.1 Comparison between the S1 and S2 extraction strategies with an extraction and simulation report on a 32CPUs of 32 × 2.10 GHz server

			Extraction time	Simulation time and memory
S1	# Nodes	18,110	Meshing = 1.41 min 2D Extraction = 1.33 min 3D Extraction = 1.2 min	Elapsed = 3 min CPU = 5 min Memory = 430 MB
	# Diodes	5922		
	# Homo-junctions	2646		
	# Resistors	22,341		
S2	# Nodes	11,483	Meshing = 1.38 min 2D Extraction = 1.33 min 3D Extraction = 44 s	Elapsed = 1 min CPU = 2 min Memory = 300 MB
	# Diodes	3276		
	# Homo-junctions	1812		
	# Resistors	14,349		

6.3.1.1 Activation of the Lateral NPN Transistor

A DC simulation of the original circuit with the extracted substrate model is carried out to reproduce measurement conditions. Figure 6.17 shows the 2D layout cross section with the biasing. To emulate the "belowground" freewheeling condition, a current, IL, is extracted from the output OUT1 in order to forward bias the N1 drain to the substrate junction while keeping both transistors in off state. A current sweep from 1 pA up to 400 mA leads the low-side transistor's drain (the output node) down to a negative bias voltage of about −1.1 V. The collected current I(VDDH) is measured at the P1 isolation well (DN well), set to VDDH = 12 V and the base current I(GND) at the ground node.

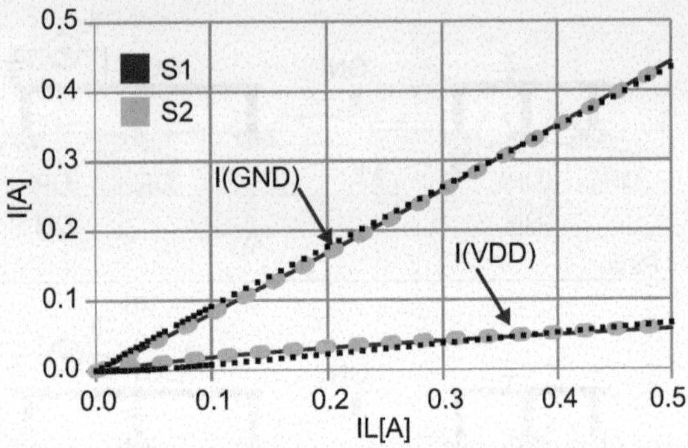

Fig. 6.18 Black dots are measurements, and continuous lines are simulation results of the S1 and S2 mesh strategies

The required technological parameters of the substrate model are calibrated following the procedure detailed in Sect. 6.1. The simulation time to compute 20 points in a DC sweep analysis with Spectre® is listed in Table 6.1 as well. Moreover, no convergence issues are encountered in any of the simulations. The high-voltage driver substrate model is extracted and simulated in a few minutes.

Finally, the measured and simulated parasitic substrate current caused by the activation of the substrate parasitic NPN is reported in Fig. 6.18. Both base and collector currents of the substrate distributed NPN bipolar transistor are plotted on linear scales as a function of the emitter current IL (note that IL = I(VDDH) + I(GND)). The coupling currents predicted by the simulations are found to be in agreement with the measurements from low to high levels of injected currents. As shown in Fig. 6.18, the S2 meshing strategy produces results identical to the S1, using only 50% of simulation time. When the power device is switching so as to generate a belowground current injection of 500 mA, 440 mA flows through the ground node, while 60 mA flows through the VDDH node that is coupled via the substrate parasitic lateral NPN transistor. This generates an undesirable excess of current consumption at the VDDH power supply.

6.3.2 *Full H-Bridge Driver*

As discussed in Sect. 2.4.1.1, during the freewheeling phase, the H-Bridge output voltages (OUT1 or OUT2 pins in Fig. 6.19b) go below ground or above supply voltage levels generating reverse currents in the circuit. Therefore, the H-Bridge circuit layout is investigated for below ground and above supply operating modes in order to evaluate the activation of substrate parasitic bipolar transistors with SPICE simulations. In this context, it is important to mention that attempts at simulating substrate couplings in a real H-Bridge IC were investigated only in [7] through extensive TCAD simulations.

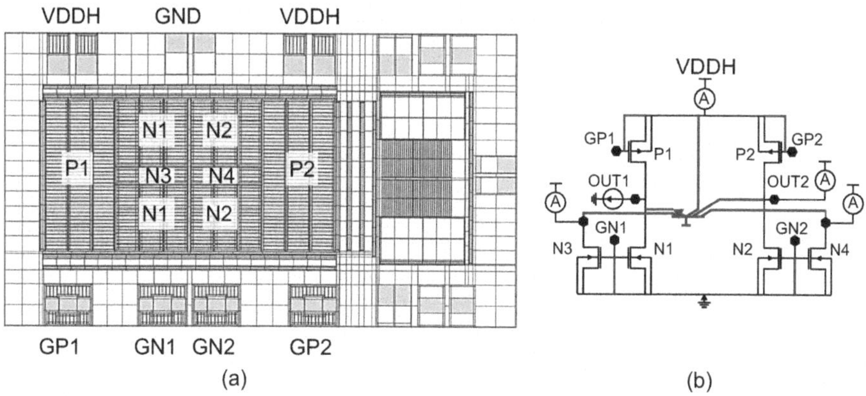

Fig. 6.19 H-Bridge layout view after meshing (**a**). Setup for OUT1 below ground measurement and simulation (**b**)

Table 6.2 Extraction and simulation report for the H-Bridge on a 32CPU of 32 × 2.10 GHz server. Simulations include both OUT1 below ground and above supply configurations for the computation of 136 DC steps

		Extraction time	Simulation time and memory
# Nodes	57,703	Meshing = 2.56 min	Elapsed = 24.3 min
# Diodes	13,451	2D Extraction = 2.16 min	CPU = 34.5 min
# Homojunctions	10,870	3D Extraction = 4.56 min	
# Resistors	70,624		Memory = 1.53 GB

The H-Bridge layout view after meshing is shown in Fig. 6.19a. For the substrate model extraction, the meshing S1 strategy in Fig. 6.17a is adopted. The S1 meshing strategy takes into account both PSUB-DN and DP-DN diodes that must be extracted to simulate effects of both lateral and vertical parasitic BJTs. Indeed, two back-to-back PSUB-DN diodes emulate a lateral NPN, while two front-to-front DP-DN and DN-PSUB diodes emulate a vertical PNP transistor.

The substrate model extraction time with the number of nodes and elements is detailed in Table 6.2. Compared to the previous extraction data summarized in Table 6.1, here the number of nodes is higher due to the complexity of the H-Bridge layout, which involves the extraction of four DMOS transistors and the entire substrate with contacts.

6.3.2.1 Output Below Ground

According to the H-Bridge operating modes, OUT1 is driven below ground either during forward-freewheeling or forward-reverse operating modes [4]. In Fig. 6.19b, a reverse current up to 1 A is injected at the OUT1 pin of the H-Bridge. Thereby, N1 drain is reverse biased, and currents coupled through the substrate NPNs are

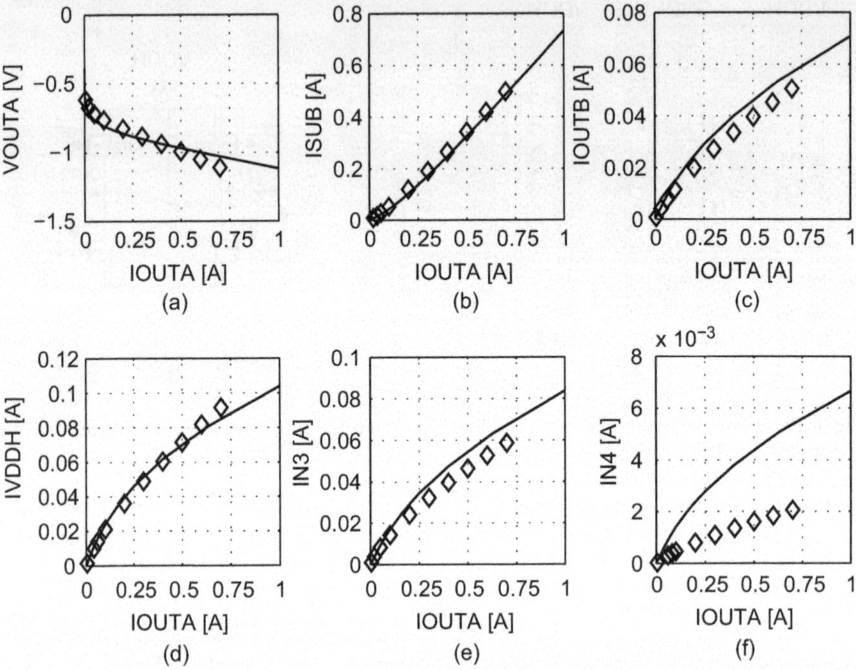

Fig. 6.20 OUT1 below ground. Simulated (lines) and measured (dots) collected currents at four different locations

measured at OUT2 pin biased at 6 V, at VDDH pin biased at 12 V, and at N3 and N4 drains biased at 6 V. This bias configuration reproduces the forward-freewheeling operating mode. NDMOS transistors N3 and N4 are in parallel with N1 and N2 and with a width significantly smaller than the power switches. These act as low-side current sensors. A comparison between the predicted and measured currents is shown in Fig. 6.20.

Figure 6.20a shows the ratio OUT1 voltage over the injected current, while the others (Fig. 6.20b–f) represent the currents flowing in the substrate and collected at each location. A good agreement is found between the experimental measurements and simulations. It should be noted that most of the current flows into the substrate, which is the base of the lateral multi-collector NPN.

The remaining current is then mainly collected by OUT2, VDDH, and the N3 drain. Surprisingly, the N3 drain collects as much current as OUT2 and VDDH despite its smaller collecting area. The reason is essentially that N3 is mostly surrounded by N1, and thus it has a wider collecting area. Concerning the current collected at the N4 drain, the agreement between simulated and measured currents is less accurate, which could be a consequence of either an inaccurate calibration of the technology model parameters or the coarse meshing used to extract the substrate model.

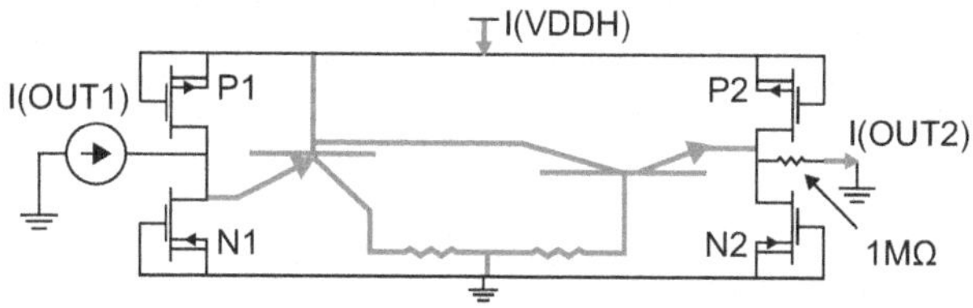

Fig. 6.21 H-Bridge circuit configuration for simulation and measurement of the OUT1 above supply condition

6.3.2.2 Output Above Supply

The H-Bridge circuit configuration for simulation and measurements of the OUT1 above supply condition is shown in Fig. 6.21. A current up to 700 mA is injected at the OUT1 pin to drive the drain of P1 above the supply voltage VDDH. In this configuration, majority carriers are injected into the substrate through the activation of the parasitic vertical PNP. Measurements and simulations of OUT1 voltage and VDDH current as a function of the injected current at the OUT1 pin are shown in Fig. 6.22a, b.

Since substrate de-biasing cannot be measured directly on the chip, it is measured indirectly through the voltage at OUT2 pin. During the measurement, OUT2 is pulled down to the ground with a $R = 1\,\text{M}\Omega$ resistor. This avoids the possibility of triggering a latch-up caused by the parasitic PNPN structure as shown in Fig. 6.21.

When the substrate voltage level crosses the threshold of the N2 substrate diode (the base-emitter voltage of the lateral NPN), the DC current flowing into the pulldown resistor increases and so for the OUT2 voltage. Therefore, the voltage at OUT2 is an estimation of the substrate voltage shifting beneath the N2 isolation n-well. The pulldown resistor current is plotted in Fig. 6.22c. Supposing that the N2 substrate diode has a threshold of $V_t \simeq 0.7$, an injected current of 300 mA generates an average substrate voltage shift below N2 of about 1.7 V. As expected, the relatively high substrate potential shift is related to the gain of the vertical parasitic PNP ($\beta_{\text{max}} \sim 15$) as shown in Fig. 6.22d.

6.3.2.3 H-Bridge Layout Implementation

The H-Bridge "belowground" or "above supply" electrical conditions are discussed with the purpose of layout optimization based on a NDMOS and PDMOS reverse current analysis.

The designed H-Bridge includes two n-type low-side NDMOSs (N1 and N2) and two lateral high-side PDMOSs (P1 and P2). The four transistors are placed in the chip layout in the order P1, N1, N2, and P2 as shown in Fig. 6.23.

During the OUT1 belowground condition, minority carriers are injected into the substrate at the N1 drain. Minority carriers which propagate leftward with respect

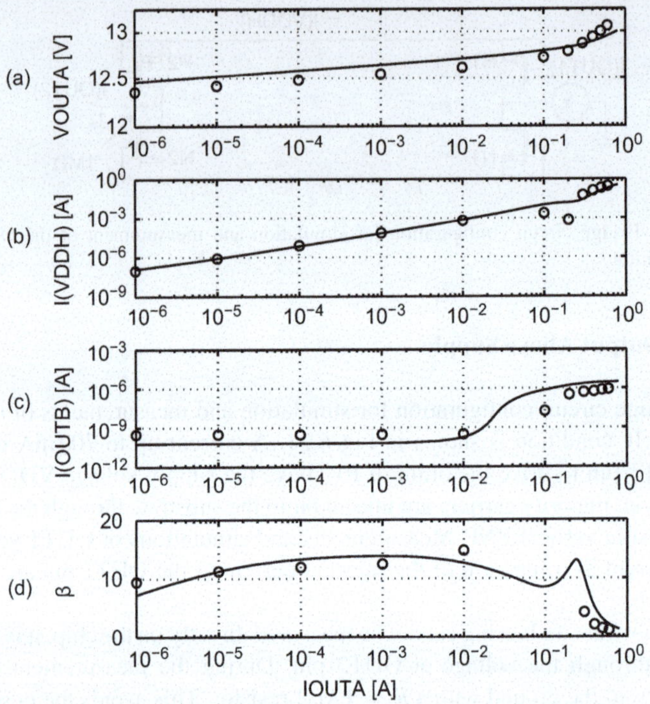

Fig. 6.22 OUT1 current-voltage characteristics during the above supply condition as a function of the injected current. Dots are measurements and lines are SPICE simulations

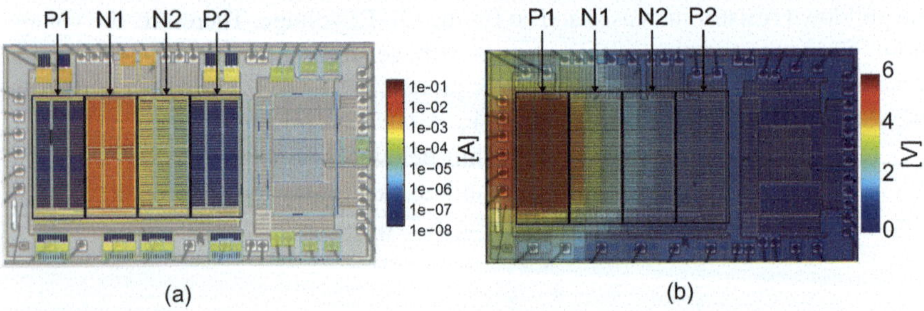

Fig. 6.23 Substrate current (**a**) and voltage distribution (**b**) for a reverse current of IOUT1 = ±0.5 A. Simulation results are superimposed on the die picture of the H-Bridge

to N1 are collected by the P1 n-well that is 530 μm wide, while minority carriers which propagate rightward are collected first by the N2 n-well (530 μm wide) and then by P2 n-well (530 μm wide) which is biased at VDDH = 12 V. Therefore, P1, N2, and P2 n-wells act as protections for N1, thus making an additional lateral n-type guard ring superfluous.

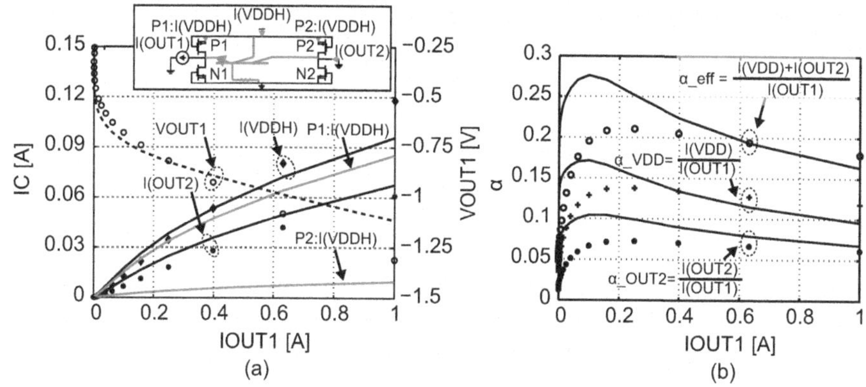

Fig. 6.24 (**a**) Measurement results of lateral NPN at $T = 80\,^\circ$C. Dots are measurement and lines are simulations. The partition of I(VDDH) current through P1 and P2 n-wells is also reported. (**b**) Measurement results of lateral NPN couplings at $T = 80\,^\circ$C. Dots are measurements and lines are simulations

This is confirmed by measurements shown in Fig. 6.24. Since substrate coupling effects are more pronounced at higher temperatures, the chip is kept at $T = 80\,^\circ$C during measurements. The substrate current is coupled with VDDH through the P1 and P2 n-wells ($I(\text{VDDH}) = I(P1 : \text{VDDH}) + I(P2 : \text{VDDH})$). Although it is not possible to measure the individual P1 and P2 contributions, simulations show that most of the currents passing through VDDH are collected by the P1 n-well since it is adjacent to the injector N1. The current collected by P2 is much lower as is mostly collected by N2 (I(OUT2) current) placed between the injector N1 and P2. The substrate current distribution issued from simulations is also plotted in Fig. 6.23a for an injected current of 0.5 A. Simulations are superimposed on the die picture of the H-Bridge.

In practice, for a reverse current at OUT1 of 0.5 A, 35 mA ($\alpha_{\text{OUT2}} = 0.07$) are coupled with OUT2 and 65 mA with VDDH ($\alpha_{\text{VDD}} = 0.13$). The maximum value for α_{eff} is in the range of 0.2–0.3 (see Fig. 6.24b). This substrate current, which generates an undesirable excess of current consumption at VDDH, must be taken into account for the design and optimization of an on-chip or off-chip voltage regulator.

Simulation is also able to capture substrate shifting during the OUT1 above supply condition for an injected current equal to 0.5 A. The substrate voltage distribution is showed in Fig. 6.23b. The substrate voltage below N2 is between 0 and 1.5 V and confirms a DC current flowing into the pulldown resistor as discussed in Sect. 6.3.2.2.

6.3.3 Low-Side NDMOS: Coupling with ESD Protection

Reverse currents due to the activation of the lateral NPN spread across the entire substrate. Consequently, reverse-biased diodes employed as ESD clamping structures unavoidably collect part of the substrate current. Figure 6.23a shows the current collected by each ESD protections in the H-Bridge during OUT1 "belowground" condition.

To highlight the coupling of NDMOS N1 with the ESD diode protection placed in the gate pad, the low-side NDMOS is isolated and characterized under reverse current. The setup for measurement and simulations is shown in Fig. 6.25a. As usual, a reverse current of up to 1 A is drawn from the NDMOS drain to determine the current-voltage (*I–V*) characteristic of the NDMOS with the drain pulled below ground. A pulldown resistor of 50 kΩ is connected between the NDMOS gate and source terminals in parallel with the ESD diode. An additional current source IREF is used to bias the gate GN1 of the NDMOS transistor to negative values. The *IV* characteristics are measured under two different bias conditions of the gate voltage (see Fig.6.25a):

- a current IREF equal to 1 nA is drawn from the NDMOS gate;
- a current IREF equal to 10 μA is drawn from the NDMOS gate.

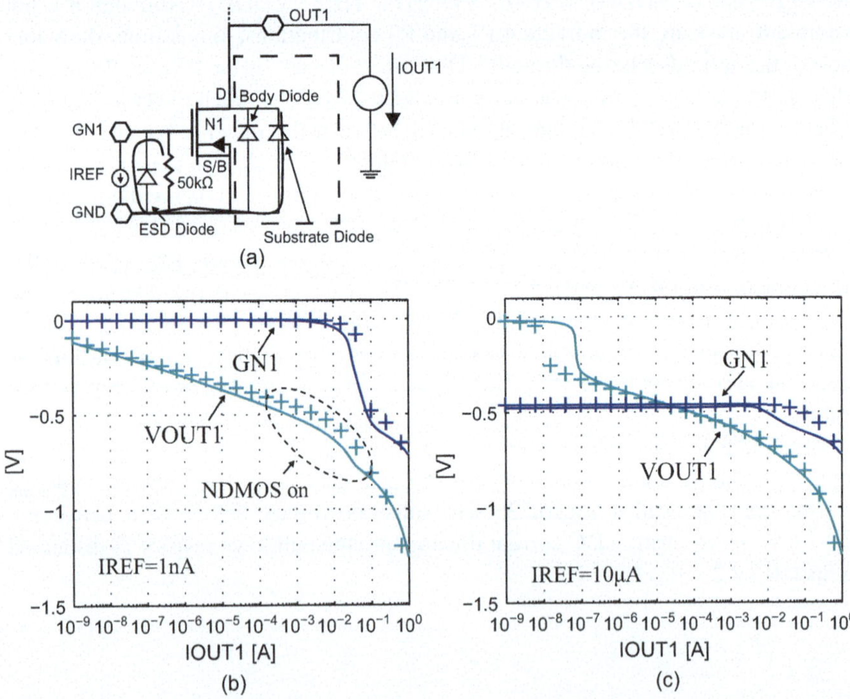

Fig. 6.25 Coupling of NDMOS reverse current with input ESD protection. Setup for simulations and measurements (**a**) with results of *I–V* characterization with IREF = 1 nA (**b**) and IREF = 1 μA (**c**). Dots are measurements and continuous lines are simulations

Simulations and measurement results of both cases are shown in Fig. 6.23b and c. In both cases the gate driving voltage *V*(GN1) is proportional to the current IREF for low levels of injected currents. For higher reverse drain current, *V*(GN1) is no longer proportional to the IREF current and becomes more negative.

With IREF = 1 nA, the gate voltage is initially equal to 0 V and drops to −0.6 V for higher reverse currents. A minor deviation is observed between the measured and simulated NDMOS drain voltage OUT1 close to the transition of the gate voltage as shown in Fig. 6.25b. The reason comes from the reverse current flowing through the NDMOS channel that is not accurately predicted by the compact model of the NDMOS.

With IREF = 10 μA, the gate voltage is biased at −0.5 V, close to the threshold voltage of the NDMOS. Thus, the NDMOS is almost totally off during the negative sweep of OUT1. In this case an excellent agreement is observed between measurements and simulations (see Fig. 6.25c). In the I-V characteristic, the OUT1 voltage starts at zero for low reverse currents and becomes negative shortly afterward. This is because part of the IREF current is injected into the substrate through the ESD diode, which initially acts as an emitter. Then, since the OUT1 reverse current increases, the ESD diode acts a collector, as confirmed by the negative transition of OUT1 voltage. This example shows how the leakage in ESD protections is increased in the case of substrate couplings.

6.3.4 Rotor Coil Driver

In this section, the case study of an integrated rotor coil driver is examined to show latch-up risk related to the use of grounded deep n-wells as suppressors of substrate noise couplings [5]. A simplified schematic view of the rotor coil driver is shown in Fig. 6.26a. A pulse-width-modulated (PWM) high-side power switch and an integrated low-side freewheeling diode control the current in the inductive load reproducing the rotor excitation coil of a car alternator. The low-side freewheeling diode is the body diode of a NDMOS transistor.

The driver is used in voltage regulators for automotive alternators [1]. A detailed description of the driver design and challenges is provided in [6]. This circuit is typically placed close to the engine in a car, and thus it has been designed to fulfill all automotive requirements as described in Chap. 2. Specifically, two very important and challenging automotive aspects are addressed in the circuit:

- **DPI applied to VBAT**. The DPI evaluation is carried out at the VBAT "global" pin to assess the robustness of the circuit during the High frequency (HF) power injection. Grounded deep n-type rings are typically used to provide a low impedance return path to the ground for substrate noise currents. The depletion capacitance of grounded isolation n-well rings is used to create capacitors deep into the substrate. Moreover, this is best accomplished with deep n-wells. This advantage is mainly due to deep n-well's large area and depth that can collect

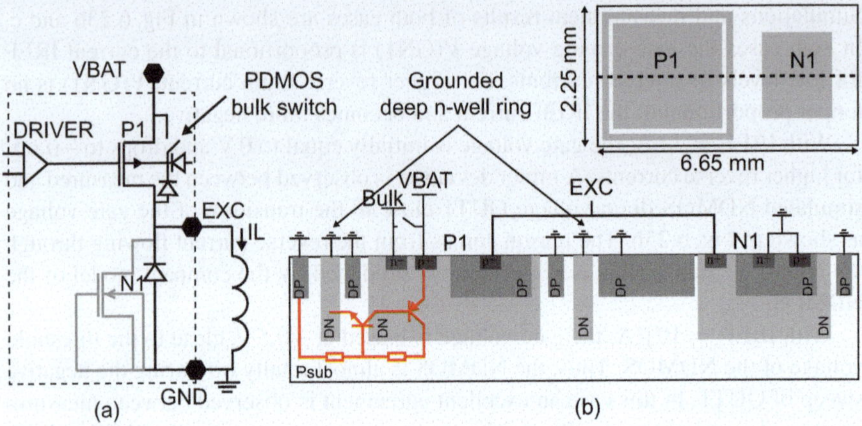

Fig. 6.26 Simplified rotor coil driver schematic (**a**) and layout (**b**)

more charges from the substrate. The PDMOS is thus enclosed by an additional grounded deep n-well as shown in the layout and its cross-sectional view in Fig. 6.26b.

- **Reverse battery protection**. Under normal operating conditions, the driver is directly supplied by the positive car battery voltage VBAT. Accidental inversion of the battery polarity can occur when the battery is installed backward as, for example, during car services. However, the reverse battery voltage is clamped to −3.2 V [6]. Indeed, the series of two power diodes built in the rectifier bridge of the alternator are forward biased under reverse battery conditions. In order to also safely protect the circuit from large currents, a bulk driver circuit is implemented to control the PDMOS n-well and to avoid the conduction of the drain to bulk diode in the case of a reverse battery fault condition. The bulk switch shown in Fig. 6.26a serves this purpose.

However, the PDMOS bulk driver circuit, in conjunction with the grounded deep n-well, can affect the circuit functionality and lead to the activation of a SCR structure; see Fig. 6.26b. To avoid the activation of the parasitic vertical PNP device comprising the source to body diode of P1 and the DN-PSUB diode, special attention must be paid to the power-up sequencing of the driver. At power up, the high impedance of the reverse battery protection switch may result in a large time constant on the P1 bulk node activating the p+/DNWELL/PSUB vertical PNP. The consequent activation of the lateral NPN (DN/PSUB/DN) triggers the latch-up.

Here, this specific situation is simulated taking into account the driver substrate model. The substrate extraction is carried out in accordance with the basic procedure referred for S1 in Sect. 6.3.1 but with the addition of the source body diode of transistor P1. This diode is required for the extraction of the vertical PNP, which results from the two front-to-front connected (p+)-DN and DN-PSUB diodes. The

Table 6.3 Extraction and simulation report (160 DC points) for the test structure on a 32CPU of 32×2.10 GHz server

		Extraction time	Simulation time and memory
# Nodes	14490	Meshing = 8.86 min	Elapsed = 4.1 min
# Diodes	4106	2D Extraction = 9.5 min	CPU = 7.71 min
# Homojunctions	2314	3D Extraction = 1.06 min	
# Resistors	18099		Memory = 410 MB

substrate extraction time with the numbers of nodes and elements is detailed in the Table 6.3.

As shown in Table 6.3, whereas the number of extracted nodes is of the same order of magnitude as in the previous cases listed in Tables 6.1 and 6.2, meshing and extraction time are longer because of the larger area of the chip ($\approx$15 mm^2). The layout view with meshing is shown in Fig. 6.28a.

6.3.4.1 Latch-Up During the Driver Power-Up Sequence

A DC analysis is performed prior to a transient analysis to investigate the substrate current and the substrate voltage distribution when the SCR structure is intentionally activated. The transient analysis is then simulated to check if the latch-up is also triggered during a fast power-up process.

6.3.4.2 DC Simulation

By applying a potential difference between the VBAT and the bulk of transistor P1, the emitter-base region of the vertical PNP BJT is forward biased. Unlike the "above supply condition" described in Sect. 6.3.2.2, the vertical PNP transistor is activated when the PDMOS bulk is pulled down relatively to the supply voltage VBAT biased at 12 V. In the former case, the PNP emitter is the PDMOS drift region related to its drain, but in the latter the emitter is the $p+$ region corresponding to the PDMOS source.

DC simulation results are presented in Figs. 6.27 and 6.28. The simulation time needed to compute 160 points in a DC sweep analysis with Spectre® is listed in the Table 6.3 as well. Figure 6.27 shows current through the grounded deep n-well ring, P1 bulk, VBAT, and substrate nodes over the base-emitter voltage applied between VBAT and P1 bulk pins (VBE of the vertical PNP transistor).

When the base-emitter voltage VBE of the vertical PNP transistor crosses 0.7 V, the current through VBAT suddenly increases and flows into the deep n-well grounded ring and substrate contacts. The high current flowing in the n-well grounded ring indicates that the lateral NPN transistor is also activated. This corresponds to the activation of a SCR parasitic structure. The current generated by

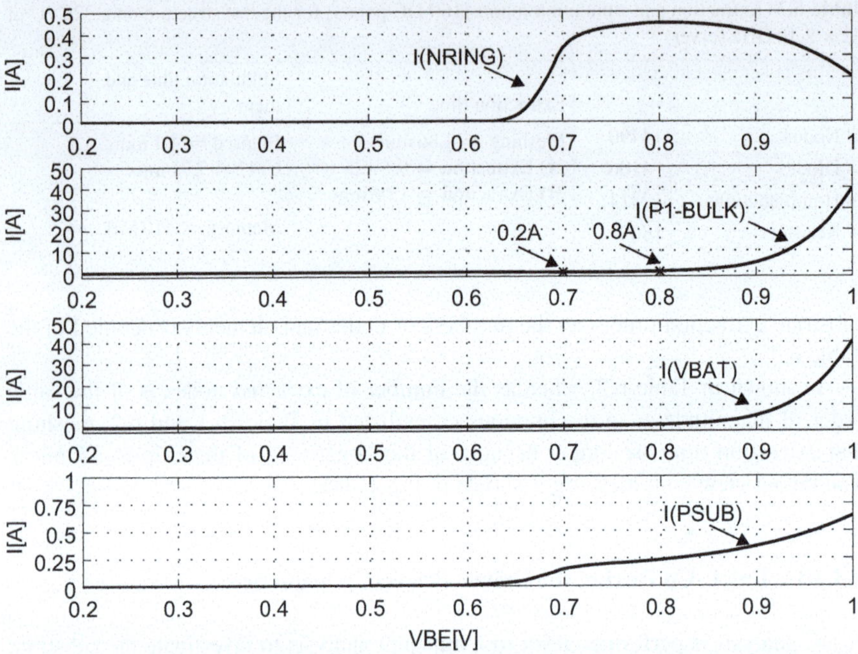

Fig. 6.27 DC simulations of substrate current through driver pins with P1 bulk pulled low relative to VBAT

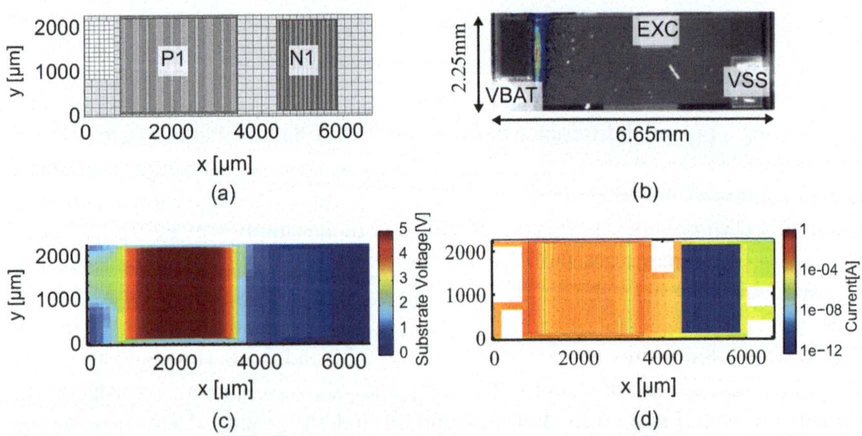

Fig. 6.28 Chip layout view with meshing (**a**). EMMI picture showing the latch-up path (**b**). Simulated substrate voltage (**c**) and current (**d**) distributions with VBE = 0.7 V

the PNP transistor causes voltage drops in the substrate. This leads to the activation of the lateral NPN transistor whose collector current sustains the base current of the vertical PNP device. This keeps the PNP transistor on, even if the external triggering current is removed. The PNP current continues to sustain the operation of the lateral NPN transistor, and this positive feedback is at the origin of the latch-up. In this situation, the lateral NPN structure is formed by the grounded deep n-well (emitter), the p-substrate connected to GND (base) and the deep n-well of P1 transistors (collector).

Thanks to simulations with the substrate model, it is also possible to find the latch-up main current paths into the substrate. These are shown in Fig. 6.28c, d. These findings are compared with the latch-up hot spot picture of the chip monitored by emission microscope (EMMI) and shown in Fig. 6.28b, where the latch-up is visible on the left-hand side of the PMOS transistor.

As can also be seen from Fig. 6.28c, the nonuniform substrate voltage potential caused by the PNP current is higher below the PDMOS structure. Then it decreases gradually to zero toward the borders of the considered chip area. The substrate potential below the grounded n-well, however, is slightly higher at the left end of P1 transistor. This is expected since areas with substrate contacts are not the same on the right and left sides of P1 structure. Therefore, since the substrate access resistance is higher on the left side of P1, the base-emitter voltage of the left-side lateral NPN is higher if compared to the NPN at the opposite end of P1. The substrate current is thus more pronounced in the left-side region as confirmed by the substrate current plot shown in Fig. 6.28d. The latch-up spot image of the EMMI picture confirms the results of the simulation.

6.3.4.3 Transient Simulations

The transient measurements are carried out with an external resistor of 10 Ω connected in series with VBAT. Then, to avoid a latch-up failure, the 10 Ω resistor limits the current in the SCR structure to

$$I_{\mathrm{LIM}} = \frac{\mathrm{VBAT} - V_T}{10} \tag{6.2}$$

where V_T is the thyristor on-state voltage. A plot with the measured VBAT voltage and current versus time is shown in Fig. 6.29a. After the initial power up, the VBAT voltage after the series resistor drops from the initial 12 V to about 3.4 V, and consequently the VBAT current is 860 mA. The circuit test bench, including the extracted substrate model, is simulated in the same electrical conditions. A fast transient on VBAT (0 to 12 V) is applied to emulate the driver power-up. Transient simulation results are plotted in Fig. 6.29b.

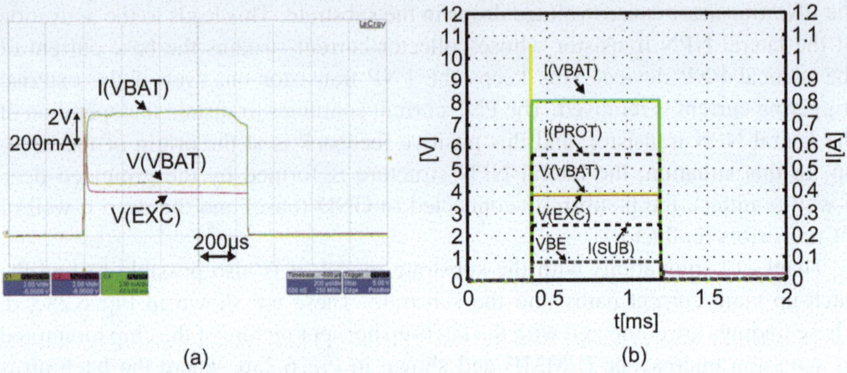

Fig. 6.29 Power-up measurements (**a**) and simulations (**b**) showing the latch-up problem. The current through VBAT is limited by an external 10 Ω resistor

A low impedance path due to the activation of a PNPN thyristor is created between VBAT and the n-well protection ring of P1. This is confirmed by the high level of current flowing in VBAT. As mentioned above, the high impedance of the reverse battery protection switch at power up (the switch is not immediately turned on) results in a large time constant on P1 bulk. This created a large transient voltage across the base-emitter junction of the p+/DNWELL/PSUB vertical PNP (VBE in Fig. 6.29b). Then, the consequent activation of the lateral NPN (DN/PSUB/DN) through P1 n-well grounded ring triggered the PNPN thyristor. In addition, the simulation captures the amount of current flowing in the substrate (*I*(SUB)) and the grounded n-well (*I*(PROT)), respectively.

While the P1 grounded n-well ring prevents substrate de-biasing noise generated by P1 from reaching other circuits, it can be fatal for the triggering of a latch-up as soon majority carriers are injected into the substrate.

6.4 Conclusions

This chapter validates the overall methodology presented so far. Simulations and measurements on a high-voltage circuit reveal consistent results and therefore confirm the validity of the method. Today, exact values of coupled current through the substrate are available only after measurements, which can generate design iterations if there are malfunctions in the IC. The simulation of such a coupling during post-layout simulations can predict substrate coupling effects, reducing the risk of having to redesign the circuit. For instance, the excess of current consumption at VDDH obtained by simulation can be taken into account in the full-chip level design for the optimization of an on-chip voltage regulator.

The activation of latch-up structure in a rotor excitation driver was also validated. The EMMI results were followed by detailed layout and substrate simulation analysis. It was found that the root cause of the latch-up is the grounded n-well used to suppress substrate noise coupling. This result is of great importance and, for the first time, validates the activation of a thyristor in a circuit with SPICE-like simulators.

References

1. L. Fanucci, G. Pasetti, P. D'Abramo, R. Serventi, F. Tinfena, P. Chassard, L. Labiste, P. Tisserand, An high voltage CMOS voltage regulator for automotive alternators with programmable functionalities and full reverse polarity capability, in *Design, Automation Test in Europe Conference Exhibition (DATE)*, pp. 526–531, March 2010
2. A. Hastings, *The Art of Analog Layout* (Prentice Hall, Upper Saddle River, 2005)
3. M. Kollmitzer, M. Olbrich, E. Barke, Analysis and modeling of minority carrier injection in deep-trench based BCD technologies, in *9th Conference on Ph.D. Research in Microelectronics and Electronics (PRIME)*, pp. 245–248, June 2013
4. B. Murari, F. Bertotti, G.A. Vignola, A. Andreini, *Smart Power ICs: Technologies and Applications* (Springer, Berlin, 1996)
5. Y. Oh, S. Lee, H. Shin, J.-S. Rieh, Trench-type deep N-well dual guard ring for the suppression of substrate noise coupling. Int. J. RF Microwave Comput. Aided Eng. **21**(1), 36–44 (2011)
6. S. Saponara, G. Pasetti, F. Tinfena, L. Fanucci, P. D'Abramo, HV-CMOS design and characterization of a smart rotor coil driver for automotive alternators. IEEE Trans. Ind. Electron. **60**(6), 2309–2317 (2013)
7. M. Schenkel, *Substrate current effects in smart power ICs*, Ph.D. thesis, ETH Zürich, Nr. 14925, 2003
8. R.K. Williams, M.E. Cornell, J.A. Harnden, MOSFET flyback-diode conduction and dV/dt effects in power ICs in low-voltage motor control applications, in *Proceedings of the 3rd International Symposium on Power Semiconductor Devices and ICs, 1991. ISPSD'91* (IEEE, Baltimore, 1991), pp. 254–257

Chapter 7
Substrate Coupling Analysis and Evaluation of Protection Strategies

7.1 Preliminary Substrate Simulations

The design of a IC requires many steps ranging from electrical design specifications to the circuit design, layout floor planning, and post-layout simulations. With the EPFL substrate model and the extraction tool introduced in Chaps. 3 and 5, IC designers can perform post-layout simulations to accurately assess the circuit robustness against parasitic substrate couplings.

In the standard circuit design flow, post-layout simulations are handled at the end of the design cycle as illustrated in the right branch of Fig. 7.1. This is the last stage before the chip fabrication and requires many iterations between design and layout engineers in order to fix parasitics-related problems. Unfortunately, the extraction of the substrate network from the final IC layout can be time-consuming, and the resulting netlist can be huge (large number of devices). Therefore the IC verification flow needs to be optimized to speed up the analysis of substrate couplings.

If the extraction of the substrate is only done on specific portions of the IC, the time needed for the extraction is significantly less, and the generated netlist is greatly reduced. For this reason, the preferred design flow for substrate current analysis is the one shown in the left branch of the flow chart in Fig. 7.1. As a result, substrate couplings are analyzed earlier and during the initial IC feasibility studies. Once the technology is fixed, substrate simulations of the preliminary layout floor plan provide fast information about substrate current distribution. From the preliminary substrate simulations, critical current paths are evaluated, and consequently, areas for placing protections are identified. If performance needs and cost requirements are not attained, technology and floor planning strategies are reevaluated. When performance needs and cost requirements are attained, the feasibility study is complete.

To summarize, the ability to perform substrate coupling analysis early in the design flow shows several decisive advantages. First, during the IC feasibility, it helps with the definition of the circuit architecture and with the choice of

P. Buccella et al., *Parasitic Substrate Coupling in High Voltage Integrated Circuits*, Analog Circuits and Signal Processing, https://doi.org/10.1007/978-3-319-74382-0_7

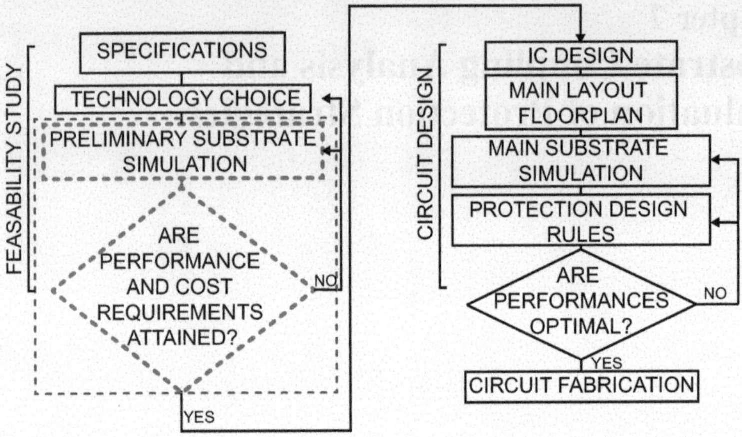

Fig. 7.1 Design flow for minimization of substrate current with early integration of substrate simulations

the technology, trading-off cost, and electrical requirements. Second, substrate simulations provide extremely important knowledge that simplifies the drawing of the IC floor plan, with the placements of all circuits and eventual protections. This approach, however, requires an expert knowledge of the circuit in order to:

1. **identify injector sources**: determine in an IC which devices, identified as injector sources, lead to the activation of parasitic NPN and/or PNP bipolar transistors;
2. **analysis of parasitic coupling paths**: run simulations of the extracted EPFL substrate model with parasitic BJT transistors for each injector device based on the preliminary layout geometries;
3. **placement of guard rings**: according to circuit voltage and current conditions, quantify substrate current though the spatial visualization of simulated data, and place optimized protections.

7.1.1 Injector Source Identification

In a HV IC, the substrate current injection sources are identified with on-chip power devices. For example, in a circuit designed to drive inductive loads, power DMOS devices are associated with the activation of substrate parasitic BJTs. When an inductive load driver, such as a Half-Bridge, is turned off, the inductor free-wheeling current causes a reverse current flow either through the LS or the HS DMOS transistor. When the reverse current flows through the LS DMOS transistor, minority carriers are injected into the substrate leading to the activation of a lateral parasitic

NPN transistor, formed by the n-type drain, the p-type substrate, and adjacent n-type regions.

Alternatively, when the reverse current flows through the HS DMOS transistor, majority carriers are injected into the substrate leading to the activation of a vertical PNP transistor, formed by the p-type source/drain, the n-type body region, and the p-type substrate. This is the case whether or not a NDMOS or a PDMOS is used as a HS driver. If sufficient current is injected into the substrate, IC failures may occur due to an excess of power consumption, substrate de-biasing, malfunction of sensitive circuits, or latch-up.

7.1.2 Substrate Extraction and Post-layout Simulation

Prior to the design of complex circuits such as Half- or Full-Bridge drivers, the extraction of the substrate model applies first to each injector source present in the IC. SPICE simulations of the extracted substrate model within circuit application voltage and current conditions allow for a quick analysis of substrate parasitic currents' geometrical distribution. The extracted substrate model is a three-dimensional schematic made of diodes, resistors, and homojunctions, where each node represents the corresponding location on the 3D layout. Thus, by monitoring the simulated node voltages and currents, the substrate voltage and current distribution is generated rapidly (see Sect. 5.5.1).

7.1.3 Placement of Guard Rings

This circuit electrical information is the input specification for designing the appropriate protection in order to reduce couplings to the required levels. Specific guard rings can be placed in the circuit layout to prevent substrate current from reaching sensitive circuits. Then, the substrate residual current can be taken into account in order to pursue the design of sensitive circuits by choosing the appropriate architecture [1]. Moreover, the substrate current geometrical distribution facilitates the IC floor plan, thus enabling the layout placement of sensitive circuits at the optimal distance from substrate current injectors.

7.2 Reverse Current in NDMOS and PDMOS

In this section, the extracted models of the LS NDMOS and HS PDMOS transistors are simulated under reverse bias condition to estimate the amount of current injected into the substrate through the activation of parasitic NPN and PNP transistors. This corresponds to an example of a preliminary simulation of a HV Half-Bridge driver undergoing reverse currents.

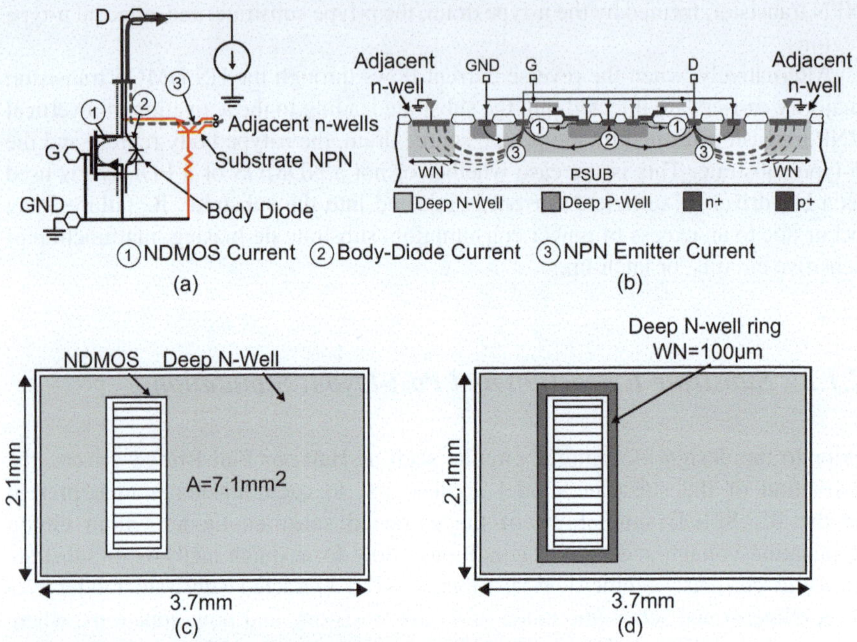

Fig. 7.2 Low-side NDMOS simulation setup (**a**) and layout cross section (**b**) with substrate parasitic devices and currents. NDMOS layout for substrate model extraction (**c**) and modified layout with 100 μm wide protection n-ring (**d**)

7.2.1 *Low-Side NDMOS: Activation of the Lateral NPN*

The NDMOS transistor schematic with substrate parasitic devices and reverse current paths is shown in Fig. 7.2a. NDMOS parasitic devices include its drain-body diode and the substrate NPN transistor, formed by the n-well (drain)/p-substrate and substrate to other n-well junctions. When the NDMOS drain drops below the substrate voltage, the NDMOS channel and the parasitic devices conduct a reverse DC current. The reverse current through the NDMOS and the body diode flows from the drain to the ground terminal. The reverse current flowing through the substrate diode, instead, flows laterally and vertically into the substrate to other n-type regions in the chip. Today, the magnitude of the total substrate current is impossible to predict during the design phase, and even the most accurate DMOS compact models which include parasitic devices exclude the effect of the parasitic lateral NPN transistor [2]. Thus, designers can only deal with general layout guidelines to reduce the risks related to minority carriers.

The layout of a NDMOS with an on-resistance of 1.2 Ω was drawn and used to benchmark substrate couplings. The NDMOS was drawn as a multi-finger structure of 108 unit transistors over an area of 530 μm × 1100 μm. The NDMOS layout cross section is shown in Fig. 7.2b. Following the technology design rules, the NDMOS is first enclosed with a p-type grounded ring, and then the resulting structure is enclosed by an additional n-type ring. The n-type ring acts as a protection against the substrate current generated by minority carriers. Indeed, when the n-well guard ring is placed very close to the injector source, it becomes the preferred sub-collector for the substrate NPN with the highest gain [3]. It collects a significantly higher percentage of substrate charges than other n-wells in the chip. Examples of n-well ring design rules are the following [4]:

- use wide n-rings, i.e., twice or triple the minimum width;
- use deep n-type wells to maximize the collecting junction depth;
- use high-doped n-wells to reduce the vertical resistance and prevent collector saturation.

However, these design rules only give qualitative information for the design of such n-type rings, whose effectiveness is difficult to estimate during chip design as the amount of substrate current is usually unknown. In order to quantify the amount of reverse current flowing through each of the NDMOS parasitic paths, the additional n-type ring surrounding the NDMOS is extended over the remaining chip area as shown in Fig. 7.2c. The n-well is covered with N+ contacts and connected to the ground and has an overall area of $7.1\,\text{mm}^2$. Such a huge n-well represents the distributed collector of the substrate NPN transistor. With this layout configuration, the total injected current and its distribution in the chip substrate are quickly estimated.

7.2.1.1 DC Simulation

The substrate model was extracted from the NDMOS layout and simulated with SPICE. The simulation was carried out by sinking a reverse DC current from the NDMOS drain in order to forward bias the parasitic junctions as shown in Fig. 7.2a. The gate was connected to the source during the simulations, since reverse recovery occurs when the NDMOS is in the off state. The curves in Fig. 7.3a show the simulated percent ratios between the reverse current flowing into the NDMOS parasitic devices and the total injected current ID. For low levels of reverse current (ID $<$ 100 mA), the current flowing in the NDMOS and through the substrate diode dominates the injected reverse current. When the NDMOS drain voltage VD becomes more and more negative, the parasitic body diode turns on, and current flows predominantly through the body diode and the substrate diode. The substrate diode current is split between substrate contacts (NPN base) and the collector according to the current gain of the distributed NPN transistor. The simulated emitter efficiency of the parasitic NPN ($\alpha_{\text{eff}} = I_C/I_E$) is shown in Fig. 7.3b. It ranges between 0.5 (for low and high injected currents) and 0.85 under the simulated conditions, so more than half of the substrate current flows toward other n-wells. Today, exact values for α_F are available only after chip fabrication or extracted from measurements of dedicated test structures.

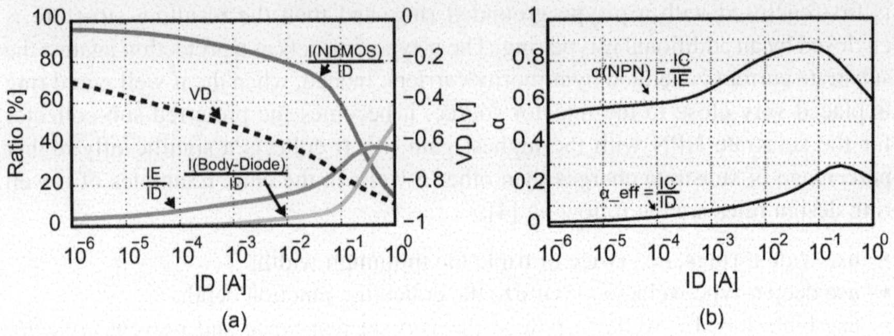

Fig. 7.3 Extracted NDMOS substrate simulation with the contribution of each current path while varying the injected current at the NDMOS drain node

Table 7.1 Substrate current distribution for a reverse current ID of 0.5 A

	$\alpha_{\max}$	Current (A)
Total substrate injected current IE	$\frac{IE}{ID} = 0.4$	0.2
Total collected current IC	$\frac{IC}{IE} = 0.85$	0.17

To design for worst case, the maximum value of α is considered. For a NDMOS reverse current of 0.5 A, the maximum substrate injected and collected currents are reported in Table 7.1.

7.2.1.2 Substrate Current Optimization

The lateral NPN collector current is distributed over the large n-well area, expected to be higher around injection areas and decrease toward the edge of the chip. This is confirmed by the color plot shown in Fig. 7.4, which reflects simulated currents from the substrate model network when the total injected current is 0.5 A. The substrate current plot is evaluated on a grid of 50 μm × 50 μm, and it enables a fast analysis for the identification of:

1. major substrate coupling current paths;
2. affected layout areas;
3. exact areas for placing protections.

The current density near the injector reaches several $\mu A/\mu m^2$. For a total injected current of 0.5 A, a n-well with an area of $50 \times 50\,\mu m^2$ close to the injector is exposed to 1.5 mA. The magnitude of collected current decreases with increased distance from the NDMOS. At a distance of 475 μm from the NDMOS, the same $50 \times 50\,\mu m^2$ area is exposed to only 4 μA. Based on simulated data, the coupled current is formulated as a function of the distance x from the NDMOS (see Fig. 7.4b), and it is fitted by

$$I_d = 1.07d^{-1.95}, \quad d[\mu m] > 50\,\mu m \tag{7.1}$$

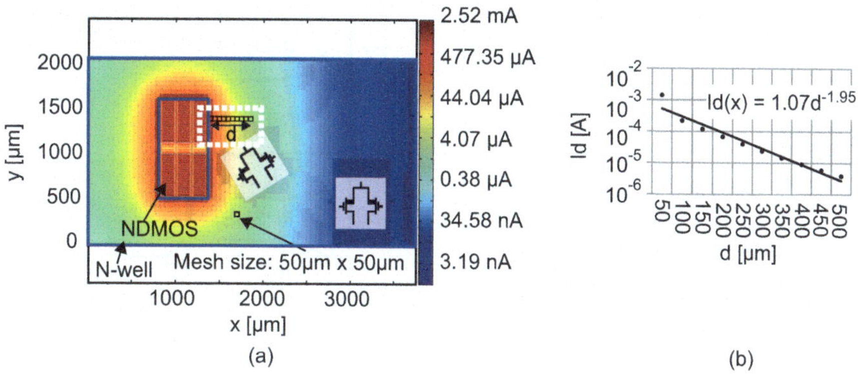

Fig. 7.4 Spatial distribution of simulated substrate current with ID = 0.5 A (**a**). Distance dependency of collected current along the x direction (**b**)

This simulation allows for the derivation of quantitative information that can be used to design unambiguously n-type rings and estimate the level of risk during the design phase. Instead of applying the approximate rule of placing a two or three times minimum width n-type ring (i.e., WN = 20 μm), the width of the n-well ring most suitable for the application is easily calculated. If, for example, a 100 μm wide protection n-ring is considered (see Fig. 7.2d), simulation results shown that 26% of the injected current is collected by the protection ring (130 mA), while 4% is spread in the remaining substrate (20 mA).

Moreover, the color plot of Fig. 7.4 allows for the optimization of IC floor planning and the placement of sensitive circuits. If two matched devices (i.e., current mirrors, differential pairs, etc.) are sensible to the substrate current, they can be placed in such a way as to receive the same amount of parasitic current in the tolerated range (see Fig. 7.4). In the same way, the architecture of a sensitive circuit such as a bandgap voltage reference can be optimized based on functionality risks related to the substrate current [1].

7.2.1.3 Transient Simulation

To investigate substrate couplings in the time domain, a negative voltage square pulse was applied to the NDMOS drain. The substrate model was extracted from the layout shown in Fig. 7.2d, which includes a 100 μm wide protection n-ring. Two reverse pulse amplitudes were set to sink reverse currents of ID = 1 mA and ID = 0.5 A, respectively. In a second simulation, a negative 100 V voltage pulse was applied to emulate the automotive ISO-7637 pulse 1 [5].

Time domain simulation results are shown in Fig. 7.5a, b. In the first situation, the reverse current of 1 mA flows mainly in the NDMOS channel (see Fig. 7.5a). The remaining current is first collected by substrate contacts during the transient rise time, but after tens of microseconds, it splits starting with the n-ring, the substrate

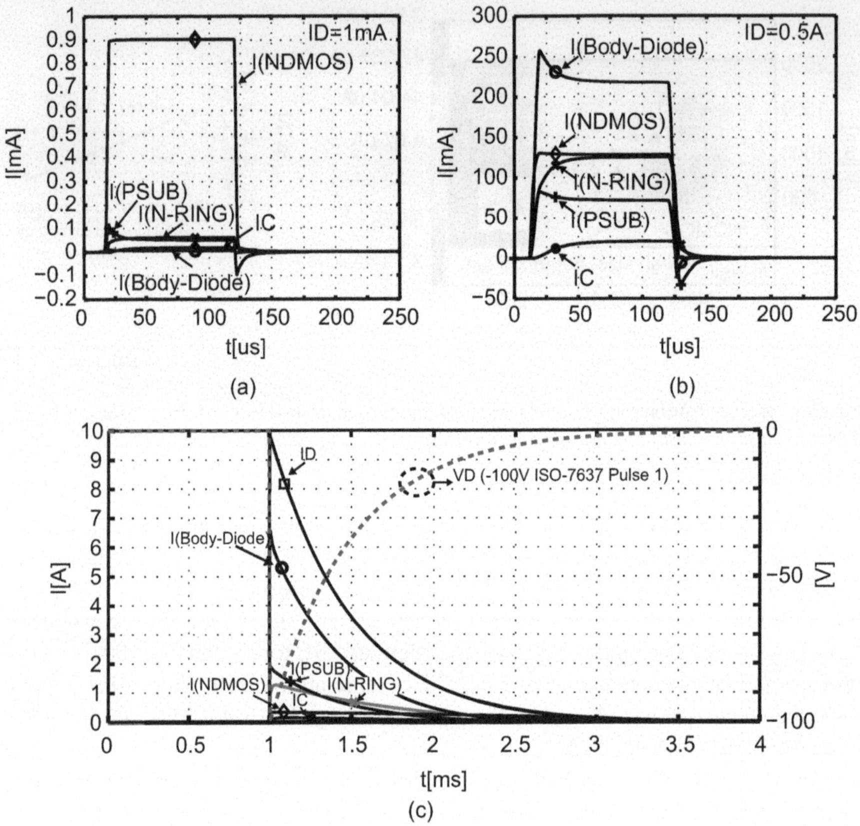

Fig. 7.5 NDMOS time domain simulations under a reverse current of 1 mA (**a**) and 0.5 A (**b**). Reverse current after a 100 V negative pulse (ISO-7637 pulse 1) (**c**)

contacts, the huge n-well, and finally the NDMOS body diode. In the second case (see Fig. 7.5b), the reverse current of 0.5 A flows from the beginning of the pulse transition through the NDMOS body diode, the NDMOS channel, the n-ring, and the substrate contacts before spreading in the huge n-well. In this particular example, after steady state is reached, 38% of the current flows into the NDMOS body diode, 26% of the reverse current flows through the NDMOS, 26% through the n-ring, and 6% through the substrate contacts, while the remaining 4% flows into the huge n-well, as already shown in Sect. 7.2.1.1.

Similarly, if a negative 100 V voltage pulse such as the ISO-7637 pulse 1 is applied to the NDMOS drain, the current flowing through the substrate can lead to circuit malfunction. The negative voltage pulse generates a reverse current ID whose peak amplitude is about 10 A when a voltage pulse generator with a series resistance of 10 Ω is considered. The distribution of the reverse current through each of the NDMOS parasitic elements is shown in Fig. 7.5c and corresponds to those resulting from the DC simulations of Fig. 7.3.

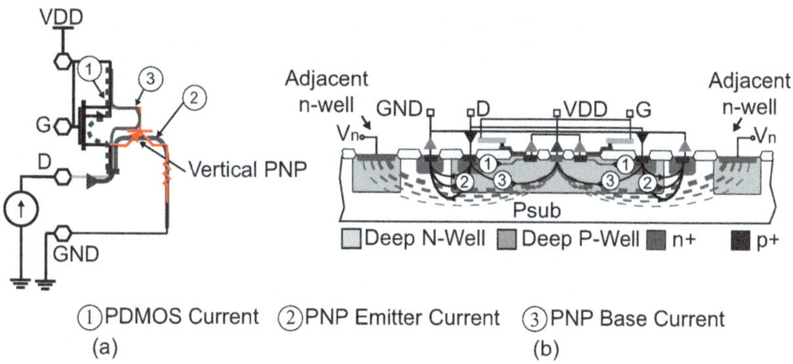

Fig. 7.6 High-side PDMOS simulation setup (**a**) and layout cross section (**b**) with substrate parasitic devices and current paths

As already discussed in Sect. 4.4, the substrate current distribution depends on the level of the reverse current also in the time domain. However, for substrate current analysis, DC simulations provide sufficient information to identify critical substrate current paths and to take the appropriate countermeasures to guarantee safe designs.

7.2.2 High-Side PDMOS: Activation of the Vertical PNP

Similarly to the LS NDMOS, when the hHS PDMOS drain is driven above the supply voltage level, the parasitic devices become forward biased and conduct a reverse current. Figure 7.6a shows the PDMOS transistor schematic with parasitic devices. The reverse current is split between the PDMOS channel and the substrate PNP transistor. The current through the PDMOS flows from the drain to the power supply, while the PNP current flows toward the power supply (via the PNP base) and into the substrate (via the PNP collector), thus generating local substrate de-biasing.

Excessive substrate potential shifts can forward bias adjacent n-wells to substrate junctions and hence lateral NPN substrate BJTs which can even cause latch-up problems. The reduction of the PNP gain is the most obvious and effective way to reduce substrate de-biasing. However, the vertical PNP transistor gain is technology dependent and not in the control of the designer.

The cross section of the PDMOS layout is showed in Fig. 7.6b. The considered PDMOS layout size is the same as that of the NDMOS ($530\,\mu m \times 1100\,\mu m$) with a switch on-resistance of $2.5\,\Omega$. General rules provide three ways to limit substrate de-biasing and avoid forward biasing the lateral NPNs BJT [4]:

- increase the substrate contact area;
- increase the spacing between n-wells and the injecting PDMOS;
- bias the n-wells to a positive voltage.

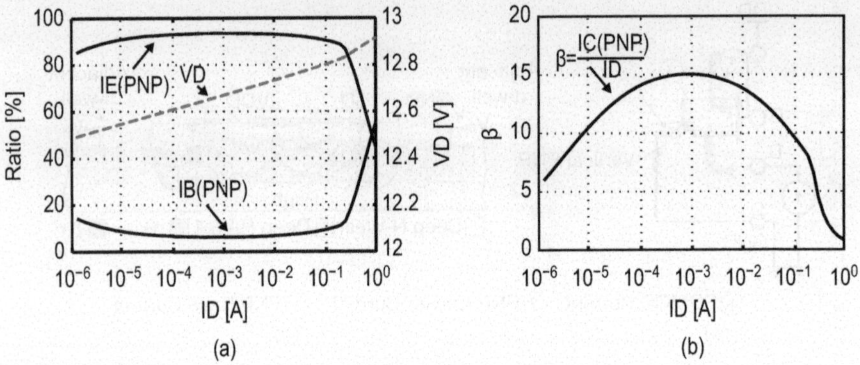

Fig. 7.7 Extracted PDMOS substrate simulation with the contribution of each current path when varying the injected current at the PDMOS drain (**a**). Simulated PNP current gain β (**b**)

In order to determine the exact spacing between n-wells and the injecting PDMOS, the PDMOS layout was enclosed by substrate contacts covering the entire remaining chip area. This layout configuration is the optimal solution in order to provide a fast estimation of the total injected current and substrate de-biasing distribution in the chip. As a result, the substrate voltage distribution helps to identify areas where other n-wells (corresponding to isolation n-wells of sensitive circuits) can be placed in the layout with the appropriate bias voltage to prevent the activation of PNPN parasitic structures equivalent to a thyristor (or SCR).

7.2.2.1 DC Simulation

The simulation on the extracted substrate model was carried out by applying a reverse current to the PDMOS drain as shown in Fig. 7.6a, with the gate connected to the source biased at VDD = 12 V. The percent ratios between the total reverse currents flowing into the PDMOS drain and the total injected current are shown in Fig. 7.7a. Unlike the NDMOS case, no reverse current flows in the PDMOS channel. This is because of the higher threshold voltage, $V_t \simeq 1$ V, of the PDMOS. The current flows predominantly through the emitter of the parasitic vertical PNP transistor and consequently into the substrate. For higher levels of reverse current, part of the current flows also through the base of the vertical PNP transistor to VDD.

7.2.2.2 Substrate De-biasing

The substrate voltage distribution is shown in Fig. 7.8 for an injected current of 0.5 A and at a substrate depth of 10 μm. The relatively high substrate potential shift is related to the high gain of the vertical parasitic PNP ($\beta_{\max} \sim 15$) as shown in Fig.7.7b. The substrate voltage distribution, as shown in Fig. 7.8, represents the best-case scenario, with the substrate contacts covering the remaining chip area.

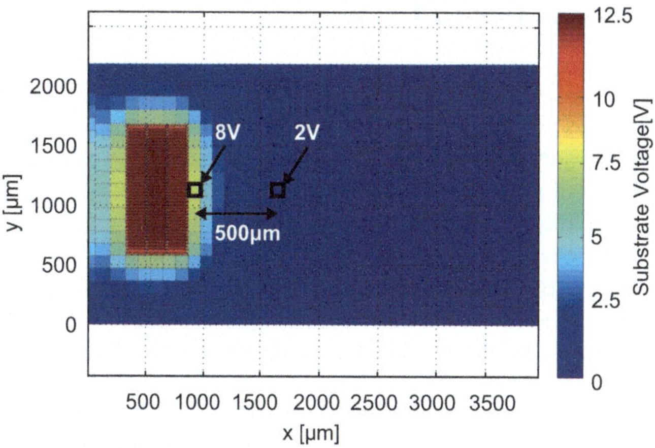

Fig. 7.8 High-side PDMOS two-dimensional substrate voltage distribution for a total injected current of 0.5 A at a depth of 10 µm in the substrate

The substrate potential beneath the n-well rises to almost 12 V because of the low-doped p-substrate ($N_a \simeq 10^{14}\,\text{cm}^{-3}$) and the absence of a backside contact. The substrate voltage reaches 8 V close to the PDMOS and then decreases to 2 V at a distance of 500 µm. Thus, a p-type ring at least 500 µm wide surrounding the PDMOS is needed to provide a low ohmic path for current injected into the substrate. The substrate voltage distribution overview clearly shows the minimum voltage level which should be applied to n-wells located near the PDMOS in order to avoid the triggering of critical PNPN parasitic structures.

7.3 Protection Strategies

The design of robust and reliable ICs immune to parasitic couplings requires the control of the substrate current and the design of adequate protections. Electrical and spatial isolation is achieved with the insertion of guard ring structures that are placed around the parasitic injector and around sensitive collecting regions. In the first case, the purpose of the guard ring is to reduce the injection and propagation of injected currents toward other regions. In the second case, the objective is to prevent currents from reaching the circuit of interest (the ring shields the sensitive circuit).

Substrate currents are originated either by majority or minority carrier injection activating substrate vertical PNP and/or lateral NPN transistors. These devices create distinct substrate coupling effects which subsequently result in different rules for designing protections. In both cases, the effects generated by the activation of such devices are controlled with p- and n-type guard rings.

7.3.1 Protection Against Vertical PNP

With the activation of a parasitic PNP transistor, the current flowing through the substrate generates voltage drops. The substrate voltage drop, if too large, can forward bias other junctions in the circuit and activate a SCRs structures.

The simulation of the substrate voltage distribution, as shown in Fig. 7.8, models both the gain of the vertical PNP and the total substrate resistance. The total substrate resistance is equal to the sum of the substrate resistance and the resistance of vertical contacts. Both resistances depend on technology and geometrical factors which are well tracked by the substrate model.

Substrate contacts should be placed near the injectors as much as possible, and once the layout is finished, additional substrate contact should be placed in the chip wherever possible. An additional Hole-collecting guard ring (HCGR) placed in the PNP n-type base region might reduce the current injected into the substrate [4].

7.3.2 Protection Against Lateral NPN

By providing an isolation area through the introduction of a protection between the power transistor (injector) and low-voltage devices (victims), the overall couplings can be reduced according to the electrical specifications of the design. The simplified cross section of a technology with an isolation area consisting of p- and n-type guard rings placed between injector sources and sensitive circuitry is shown in Fig. 7.9. This case represents the well-known aggressor-victim situation which occurs in a chip. Such guard rings when opportunely biased are efficient against minority carrier injection. The central n-well provides an additional and preferred collector to the parasitic lateral NPN for its proximity to the injecting area. According to their bias configuration, guard rings are classified as passive and active guard rings [6].

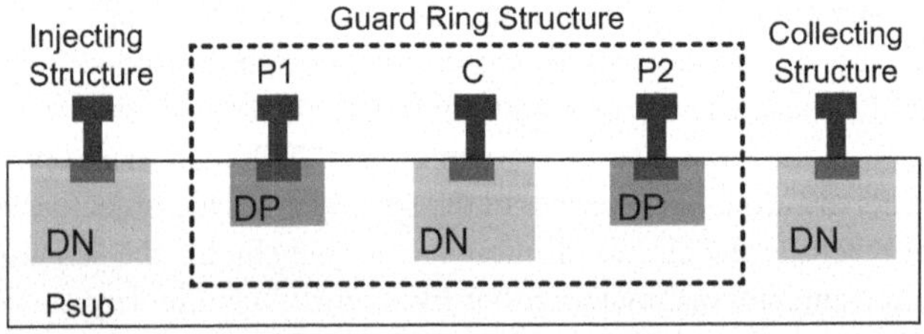

Fig. 7.9 Guard ring structure placed between the injector and the collector

7.3.2.1 Passive Guard Rings

Passive guard rings consist mainly of p-substrate contacts in conjunction with n-wells biased to a fixed voltage. This n-type guard rings are referred to as Electron-collecting guard ring (ECGR) in [4]. With reference to Fig. 7.9, p-type substrate contacts are grounded, while n-wells are connected either to the ground or to a positive supply voltage. Substrate contacts lower the substrate access resistance, thus preventing ohmic de-biasing. N-wells based on reverse-biased junctions instead collect electrons, and they can gather even more electrons if deep n-wells are used.

As a generic rule, n-type guard rings must be connected to the highest available supply voltage to drive the depletion region as deeply as possible. As discussed in Chap. 2, this biasing solution increases power dissipation caused by the activation of a lateral NPN substrate transistor. To minimize power dissipation, guard rings are sometimes grounded which makes them more susceptible to de-biasing, in the case of the simultaneous activation of a vertical NPN transistor, or less effective, due to the earlier saturation of the added collector (the collector current cannot be increased more even if the emitter current increases). However, a grounded guard ring can be supplemented by a second guard ring connected to the power supply and placed outside of the grounded guard ring. The secondary guard ring provides an additional path for substrate parasitic currents and limits the effect due to excessive substrate de-biasing if it occurs [4].

7.3.2.2 Active Guard Rings

In active guard rings, substrate contacts and n-wells are connected in such a way that they generate electrical fields in the substrate to prevent the injected minority carriers from propagating toward sensitive circuits. The active guard ring structure is similar to the passive one but has a different biasing strategy. The n-type ring is connected to the p-type contact, which is placed either near the injector or near the low-voltage area, and the resulting entity is left unconnected. The first configuration is known as the Multi-ring active analogic protection (MAAP) [7], while the second is known as the Negative feedback activated (NFA) [8]. Both structures are shown in Fig. 7.10a, b.

Whereas both isolating structures are self-triggered by the injected current in the substrate, the blocking characteristic of MAAP and NFA protections follows different processes. When the MAAP protection is activated, a negative electrical field is established around the floating p-type contact P2 connected to the n-well C. This prevents the electrons from reaching the low-voltage collector as the floating n-well C acts as the preferred collector. The unbiased guard ring reduces the gain of the lateral parasitic NPN transistor between the injector and the sensitive collector.

On the other side, in a NFA configuration, the negative electrical field is established around the floating p-type contact P1 which is closer to the injector. As the negative injected current increases, the current collected by the n-well C increases and lowers the potential below P1. This pulls down the voltage drop across the injecting junction, and thus the injected current is reduced.

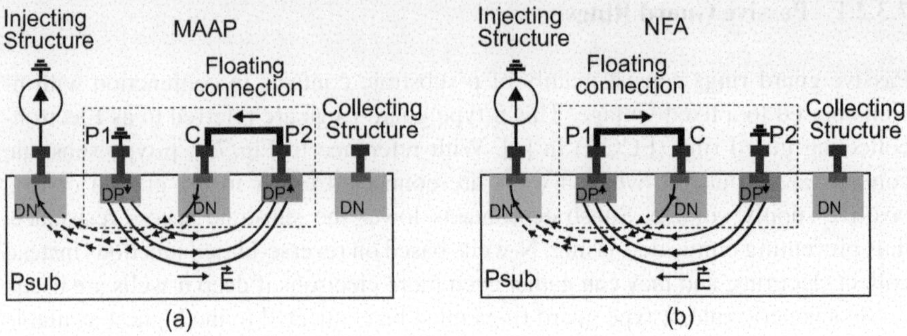

Fig. 7.10 Active guard ring cross section: MAAP (**a**) and NFA configurations (**b**)

Nevertheless, the efficiency of both protecting structures decreases when high injection phenomena are involved. Indeed, the parasitic NPN transistor between the emitter and the floating n-well C saturates for high currents. Moreover, grounded substrate contacts placed close to the protection lower the electrical field into the substrate and thus affect the protection efficiency. Both active protection structures are discussed through simulations in the next sections.

7.3.3 Guard Ring Design Parameters

The efficiency of a guard ring structure is measured as the ratio of measurable effects (i.e., substrate de-biasing, collected current) over the injected current. This ratio often represents the figure of merit of a protecting guard ring. The decrease of such ratio is a measure of the guard ring's ability to minimize the effect of injected currents.

The activation of a vertical PNP has a direct and measurable effect on the substrate voltage. Thus the substrate voltage de-biasing when compared to the injected current represents the figure of merit for the evaluation and the characterization of substrate contacts used as protection. Substrate de-biasing is difficult to quantify as it is not possible to measure it directly. An estimate is derived through the measurement of the gain α_F of the vertical PNP and the substrate resistivity [9].

The activation of a lateral NPN transistor has a direct and measurable impact on the reverse current of other n-wells in the circuit. The lateral NPN gain α_F, defined as $\alpha_F = I_C/I_E$, is a measure of the effectiveness of the guarding [10]. The lower/higher the α_F is, the better/worse the protection is. The characteristic of the α_F factor versus the distance between the collector and emitter area d provides quantitative information about the optimal distance to place sensitive circuitry from the power device in order to minimize the coupling during steady-state conduction [11].

Main protection strategies are detailed in the comparison Table 7.2 and include the main geometrical design parameters and rules. With the equivalent EPFL sub-

Table 7.2 Protection strategies with design parameters

Type of substrate current	Effect	Protection type	Design parameters	Rules	
Majority carriers (Vert. PNP)	Substrate de-biasing	Substrate contacts	Contacts area	Surround the majority carrier injecting devices with large substrate contact areas	Passive protection
Minority carriers (Lat. NPN)	Substrate currents coupling	Distance	Distance	Distance filled with substrate contacts to reduce couplings. This strategy is not efficient due to the large area consumption	
		N-well guard ring	n-Well width	Surround the minority carrier injecting device with grounded n-well areas. This solution is very efficient from low to high injection	
		Floating n-well guard ring	• n-well width • p-contacts width • p-contacts placements	Surround the minority carrier injecting devices with floating n-well areas. This solution is strongly affected by the substrate contact distribution, and the efficiency depends on the level of injection	Active protection

strate model extracted from the layout, parasitic BJT transistors are automatically detected. Then, by knowing main substrate current paths and their distribution, the efficiency of guard rings used as protection can be analyzed with SPICE simulations. In the next sections, the design and optimization of protections is done through the substrate extraction of sensible layout areas and the analysis of simulated substrate voltages and currents.

7.4 Comparative Study of Different Protection Strategies

In order to design appropriate protection structures for any given layout, it is essential to know the distribution of the substrate current. As discussed in Sect. 7.2.1, the simulation of the NDMOS substrate model gives a clear overview of substrate current distribution. The dependency of the parasitic currents on the distance from the injector in the two dimensions can then be easily visualized through substrate current distribution plot.

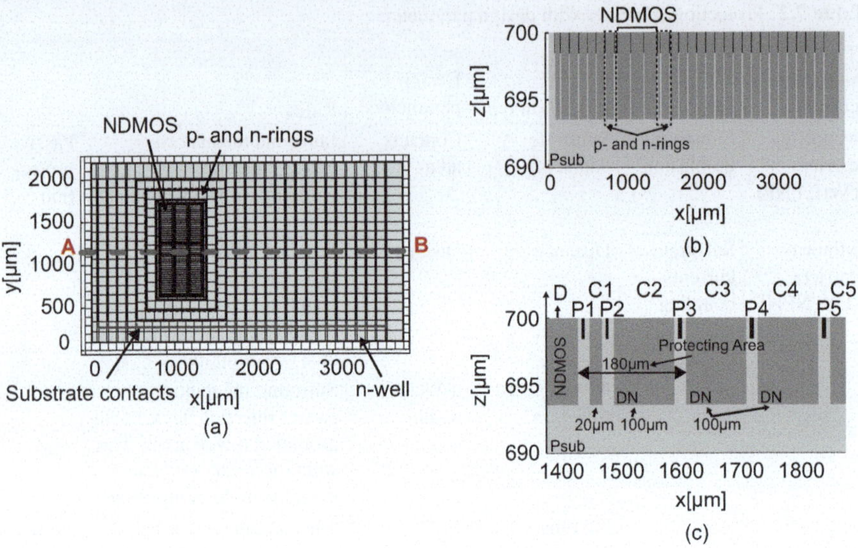

Fig. 7.11 Meshed layout of the test structure including the NDMOS and a huge n-well and substrate contacts (**a**). Layout cross section along the line AB (**b**) and with a zoom around the NDMOS (**c**)

In this section a range of protection structures is compared to one another to investigate their effectiveness against minority carriers generated by the activation of a lateral NPN transistor. All conclusions are based on simulation results of a NDMOS whose drain is the injection point when biased below ground. For the comparative study, the NDMOS is surrounded by various passive and active protecting rings. This comparative study draws on established research papers, such as in [7, 8, 12–14]. These are used for comparison with this substrate current analysis approach for deriving some key aspects for the design of optimized protection structures.

For the simulations, the layout of the NDMOS structure of Fig. 7.2 was modified to include additional substrate contacts disposed at regular intervals in the huge n-well surrounding the NDMOS as shown in Fig. 7.11a. This example is closer to a real IC and also takes into account the impact of secondary ground contacts on the protection efficiency.

The NDMOS layout cross section along the line *AB* is shown in Fig. 7.11b. A detailed picture of the structure in close proximity of the NDMOS is shown in Fig. 7.11c with substrate contacts P1–P5 and n-type guard rings C1–C5. The substrate nearby the NDMOS is grounded through the p-type contacts P1–P5. In particular, the deep n-wells C1 (WN $= 20\,\mu\text{m}$), C2 (WN $= 100\,\mu\text{m}$), and C3 (WN $= 100\,\mu\text{m}$) located, respectively, between P1 and P2, P2 and P3, and P3 and P4 substrate contacts are the n-type guard rings enclosing the NDMOS transistor. For the simulation, a reverse current of up to 10 A is applied at the NDMOS drain in order to emulate the maximum current that results when a negative pulse with an amplitude of 100 V is applied. With a NDMOS area of about 0.5 mm^2, the maximum reverse current density is 20 A mm^{-2}.

Electrical parameters of the parasitic lateral NPN transistor are compared hereafter with equal reverse current. The collected current is plotted versus the reverse NDMOS current, applied as a current source to the NDMOS drain. This approach is preferred to the classical Gummel plot measurement, where the collector current is referred to the base-emitter voltage of the bipolar transistor. Indeed, in electronic applications, the belowground condition is generated by the reverse current of an inductive load rather than a negative voltage source. Moreover, the NDMOS reverse current sets the bias point of the parasitic lateral NPN transistor, and it is more convenient for a direct comparison of the effectiveness of protections at equal reverse current.

The current measured at collectors C3, C4, and C5 over the injected current shows the effectiveness of the C1 and C2 guard rings used as passive or active protections under different configurations.
Therefore, two figures of merits were computed during simulations defined as

$$\alpha_{\text{IE}} = \frac{I_E}{I_D} \quad \text{and} \quad \alpha_{\text{NPN}} = \frac{I_C}{I_E} \tag{7.2}$$

As discussed in Sect. 7.2.1.1, only a part of the NDMOS reverse current is actually injected into the substrate. Therefore α_{IE}, defined as the ratio between the lateral NPN emitter current and the NDMOS total drain reverse current I_D, indicates the amount of the current injected into the substrate. Of this, only a part couples with other n-wells according to the parameter α_{NPN}, defined as the ratio between the collected current I_C and the NPN emitter current I_E. The absolute value of the collected current I_C is also equal to $I_C = \alpha_{\text{NPN}}\alpha_{\text{IE}}I_D$.

7.4.1 Passive Rings: Biasing Configuration

Figure 7.12 shows the cross section of a standard passive protection. The structure is a multi-collector lateral NPN transistor. The ratios α_{IE} and α_{NPN} of the collected current over the NDMOS drain current I_D are measured at different locations as well as under different biasing schemes of the two p-rings P1 and P2. Table 7.3 lists the configurations of guard rings for each case in order to reproduce results discussed in [12].

Distance Effects To show p-ring distance effects, the current I_C is first measured at C2–C5 located just 20 μm away from the injector and then at C3–C5 located 120 μm away from the injector. In both cases, the area between the injector and the collector is filled with grounded substrate contacts (C1 and/or C2 are replaced by substrate contacts). These correspond to the first two rows $d = 20\,\mu\text{m}$ *and* $d = 120\,\mu\text{m}$ of Table 7.3.

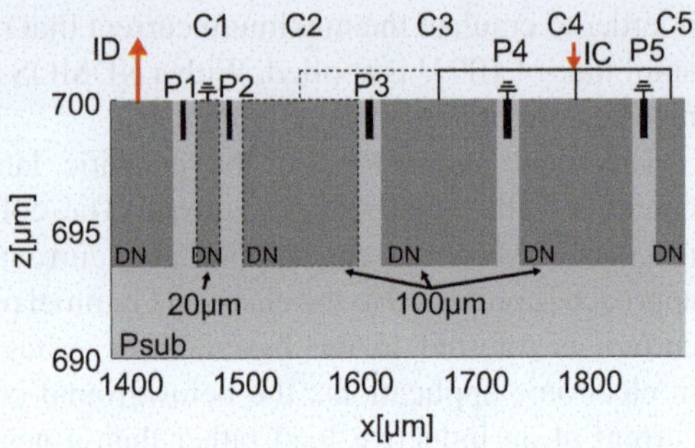

Fig. 7.12 Cross section of a passive guard ring

Table 7.3 Simulated passive guard ring configurations

Name	Substrate contacts	N-well biasing	IC
$d = 20\,\mu m$	P1–P5 grounded	C1 floating	C2–C5
$d = 120\,\mu m$	P1–P5 grounded	C1 and C2 floating	C3–C5
GG WN = 20 μm	P1–P5 grounded	C1 at +5 V	C2–C5
GG WN = 120 μm	P1–P5 grounded	C1 and C2 at +5 V	C3–C5
GF WN = 20 μm	P2 floating other grounded	C1 at +5 V	C2–C5
GF WN = 120 μm	P3 floating other grounded	C1 and C2 at +5 V	C3–C5
FG WN = 20 μm	P1 floating other grounded	C1 at +5 V	C2–C5
FG WN = 120 μm	P1 and P2 floating other grounded	C1 and C2 at +5 V	C3–C5
GF WN = 20 μm	P2 floating other floating	C1 at +5 V	C2–C5

Passive Guard Rings By design, the intermediary collecting n-rings C1 and/or C2 biased at +5 V absorb part of the injected current. They serve as passive guard rings and collect some of the substrate current before it reaches the remaining collectors. The wider the n-ring, the more current is collected. For this reason two additional configurations are discussed. In the first configuration, the collector C1 is biased at +5 V, and it is considered as a 20 μm protecting collector. In the second configuration, C1 and C2 are connected together to extend the protecting collector to 120 μm. In both configurations, all substrate contacts are grounded. These correspond to the rows *GG* in Table 7.3.

Substrate Grounding Scheme The substrate grounding scheme through p-rings P1 and P2 affects the current collected by the intermediary collector C1. As discussed in [12], the substrate contact bias scheme changes the base access resistance of the lateral multi-collector NPN transistor. Two possible grounding biases are discussed for different emitter-collector distances:

- P1 Grounded and P2 or P3 floating for an emitter-collector distance of 20 μm and 120 μm, respectively: *GF* in Table 7.3

- P1 Floating and P2 or P3 grounded for an emitter-collector distance of 20 μm and 120 μm, respectively: *FG* in Table 7.3

Effect of Multiple Ground Contacts Furthermore, since the base of the lateral NPN is the whole substrate, other substrate contacts affect the performance of the p-ring biasing scheme. Therefore, the simulation of the *GF* configuration with the collector C1 as protection is repeated with all other substrate contacts left floating. This corresponds to the configuration in the last row of Table 7.3. The simulated α_{IE}, α_{NPN}, and *IC* for each configuration are shown in Fig. 7.13. Based on the simulated data, the following conclusions are drawn:

- As shown in Fig. 7.13a, c, the amount of the injected current into the substrate I_E changes with respect to the reverse current I_D, the emitter-collector distance, the p-ring bias configuration, and the n-ring width. However, the injected current I_E is roughly the same for an equal reverse current I_D below 1 mA. This applies to all configurations. If the reverse current I_D lies in the range [1 mA, 0.1 A], the injected current is reduced by 35% for both distances and *FG* configurations with respect to the *GG* configuration. A striking difference is observed for reverse currents above 0.1 A. In the *FG* configuration, the floating P1-ring immediately next to the injecting region increases the NPN base resistance, and thus the injected current is considerably reduced.
- The ratio α_{NPN} between the collected current I_C and the injected current I_E is shown in Fig. 7.13b. Here, the behavior of the lateral multi-collector NPN transistor is strongly affected by both bias and geometrical parameters. At an equal collector-emitter distance, i.e., 20 μm, the collected current is always higher when substrate contacts are used to fill the 20 μm wide region. However, for an injected current I_E greater than 0.3 A, the collected current increases if a *FG* configuration 20 μm wide is used. In addition, the α_{NPN} parameter for *FG*, *GG*, and *GF* biasing configurations is consistent with those presented in [12]. The collected current is higher in a *FG* bias configuration with respect to both *GG* and *GF* configurations.

  In [12], it is also reported that the collected current is reduced by half when the *GF* configuration is compared to *GG*, while in this example the difference is insignificant. However, the *GF* configuration is sensitive to secondary substrate contacts which, when disconnected, lower the collected current as shown in Fig. 7.13b.
- Finally the total collected current I_C is plotted in Fig. 7.13d with respect to I_D. If, for example, the maximum reverse current is 10 A, a 120 μm wide protecting area must be used to lower the collected current to 0.1 A regardless of the biasing scheme.

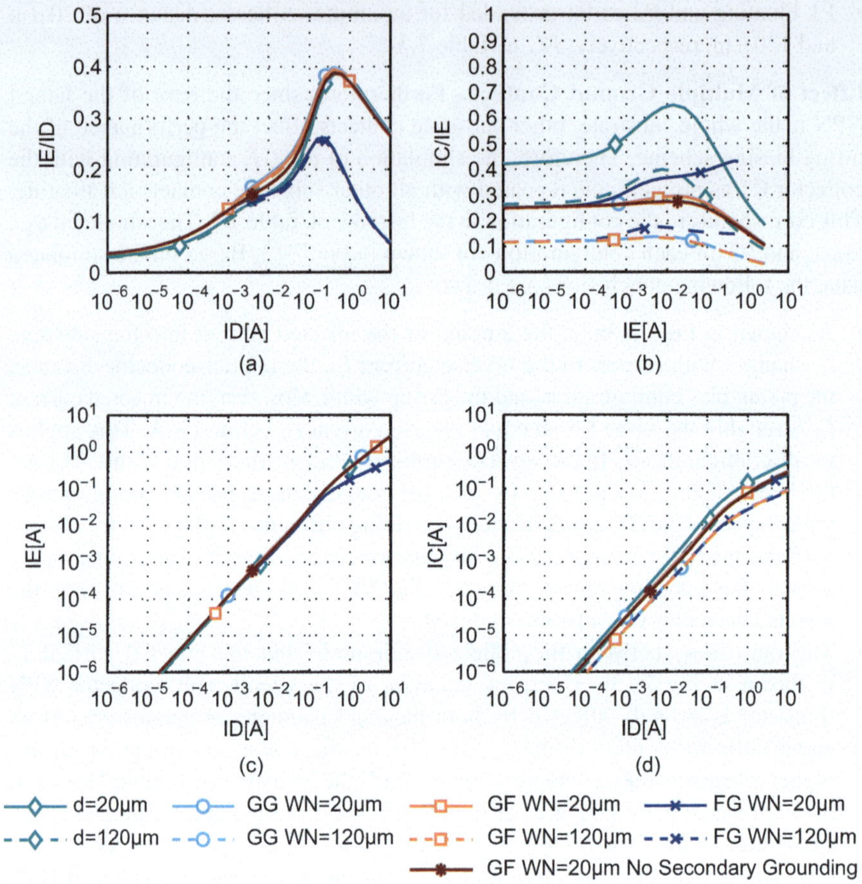

Fig. 7.13 Comparison of electrical characteristics of passive protections as a function of the reverse current I_D. Simulated injected current ratio (**a**), lateral NPN gain (**b**), total injected current (**c**), and total collected current (**d**). Different colors correspond to different guard ring structure

7.4.2 Single MAAP and NFA Active Guard Rings

By changing the electrical connection of the rings, for example, with the p-ring connected to the enclosed n-ring, new guard ring structures are designed and are known as active protections in literature.

Figure 7.14 shows two examples of active guard rings: Fig. 7.14a illustrates the cross section of a MAAP [7], while the configuration of Fig. 7.14b represents the NFA protection [8]. In the MAAP configuration, the p-ring close to the injecting point (P1) is grounded, while the second p-ring (P2) is connected to the n-ring (C1). In the NFA configuration, the p-ring close to the injecting point (P1) is instead connected to the n-ring (C1), while the second p-ring (P2) is grounded.

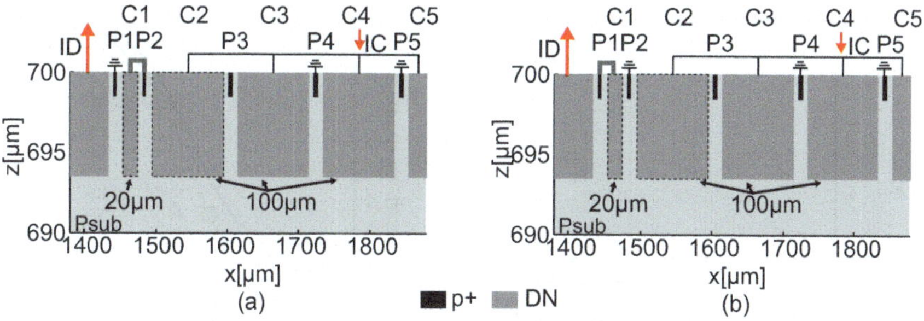

Fig. 7.14 Configuration of a MAAP (**a**) and NFA (**b**) active protection 20 μm wide

Simulations of the collected current with the NDMOS under reverse currents and enclosed by an active ring were done to show the efficiency of each protection. Moreover, the blocking efficiency of both MAAP and NFA is evaluated with a 20 μm and a 120 μm wide n-ring in the following configurations:

- MAAP 20 μm: P1 grounded and P2 connected to C1, the current is measured at collectors C2–C5 (see Fig. 7.14a);
- MAAP 120 μm: P1 grounded and P3 connected to C1 and C2 while P2 is left floating, the current is measured at collectors C3–C5;
- MAAP 120 μm_G: to show how extra substrate contacts reduce the effectiveness of the active protection, the MAAP structure with WN = 120 μm is also simulated with other substrate contacts disconnected, and the current is measured at collectors C3–C5;
- NFA 20 μm: P1 connected to C1 and P2 grounded, the current is measured at collectors C2–C5 (see Fig. 7.14b);
- NFA 120 μm: P1 connected to C1 and C2 while P2 is left floating, the current is measured at collectors C3–C5.

Simulation results are shown in Fig. 7.15. As shown in Fig. 7.15a, c, the amount of injected current into the substrate changes depending on the specific active protection type and width. Initially, the injected current follows the same behavior, but above $I_D = 1$ mA, the injected current I_E increases in the MAAP structure, while it decreases in the NFA structure. In the MAAP structure, the injected current starts to increase when the saturation point of the protection is reached. This happens when the floating connection gets closer to the injector negative voltage due to the higher current flowing in the protecting structure.

In the NFA structure, the injected current is much lower and controlled. Unlike the MAAP, the floating substrate contact close to the injection point locally changes the substrate potential, thus reducing the injected current. However, the lateral gain α_{NPN} of the active protection structures perform differently. In the MAAP, the lateral gain is quite controlled, and, before reaching the saturation level, it presents the typical *gain kink*, which is more pronounced in the 120 μm wide structure without secondary ground contacts as shown in Fig. 7.15b. The lateral gain of the NFA structure, on the other hand, is much higher.

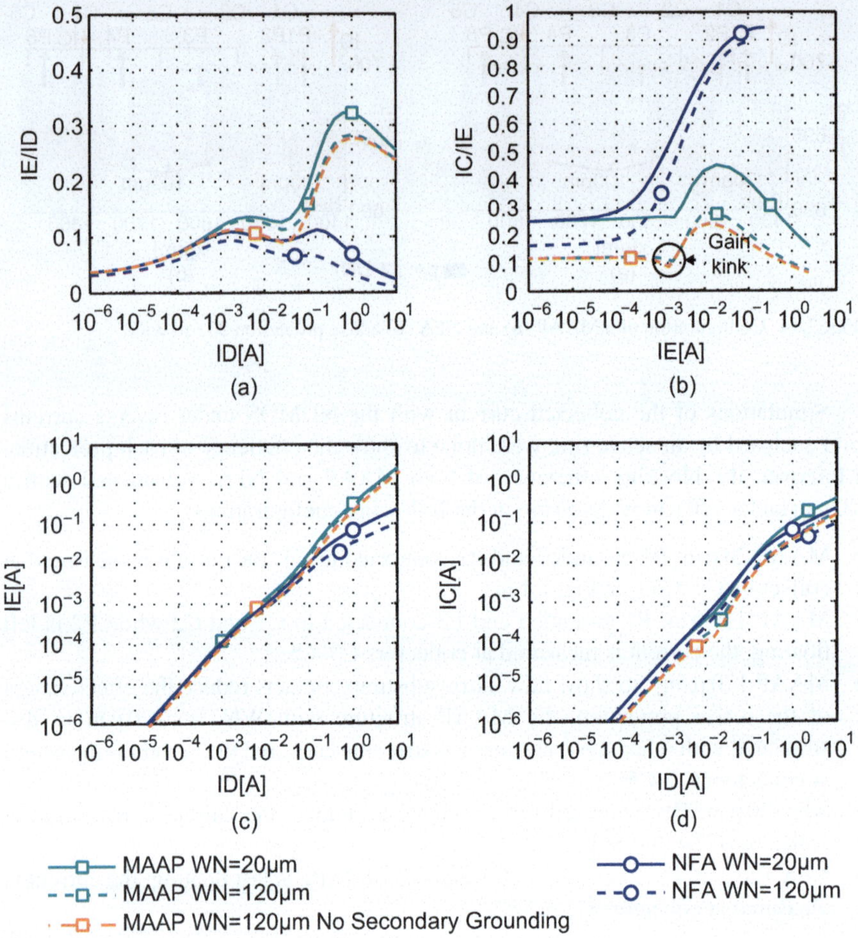

Fig. 7.15 Comparison of electrical characteristics of active protections as a function of the reverse current I_D. Simulated injected current ratio (**a**), lateral NPN gain (**b**), total injected current (**c**), and total collected current (**d**)

The substrate voltage distribution helps to explain the differences in the lateral gain levels mentioned above. As discussed in [7], the effectiveness of an active protection relies on the substrate potential profile below the protection itself. The substrate potential of the simulated structure with a MAAP and NFA with a 120 µm wide ring and an injected current $I_E = 0.1$ A is shown in Fig. 7.16. The lateral potential below the protection structure at a substrate depth of 50 µm is also plotted. Below the NDMOS, at equal injected current ($I_E = 0.1$ A), the substrate voltage is 0 V in the MAAP structure (P1 is grounded), while it is about −0.08 V in the NFA structure (P1 is floating). Below the substrate contact P3, the substrate voltage is about −0.2 V in the MAAP structure (P3 is floating), while it remains almost unchanged in the NFA structure (P3 is grounded).

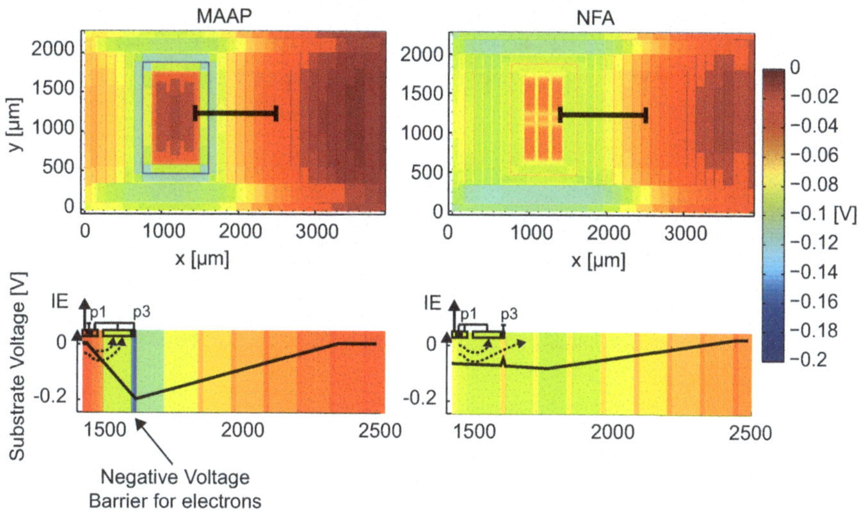

Fig. 7.16 Substrate voltage distribution with MAAP and NFA active protection (120 μm wide) with $I_E = 0.1$ A. The lateral substrate potential below the protection structures at a depth of 50 μm is also shown

With $I_E = 0.1$ A, the MAAP structure has reached saturation (see Fig. 7.15b), and the negative voltage below P3 is a barrier for the electrons. This also forces the current to flow through the substrate P1 ring. No barrier is present below the NFA structure, and hence the current flows laterally with no limitations toward other collectors. This explains the different behavior of α_{IE} and α_{NPN} in MAAP and NFA structures.

With regard to the total current collected (see Fig. 7.15d), with equal width, the MAAP is more efficient before saturation and thus at low currents. The NFA protection is instead more efficient at higher reverse currents. The enhanced performance characteristics of the NFA structure are mainly due to the reduced injected current thanks to the floating substrate condition imposed near the injector.

7.4.3 Double MAAP and NFA Active Guard Rings

To improve the effectiveness of an active protection, the use of consecutive MAAP or NFA structures is investigated. In this section, single active isolation structures (as shown in Fig. 7.14) are compared with double structures as in [8].

The configuration of an active protection with two consecutive MAAP structures is shown in Fig. 7.17a. An active protection with two consecutive NFA structures is shown in Fig. 7.17b. The collector-emitter distance is the same for both single and double configurations and equals to 180 μm.

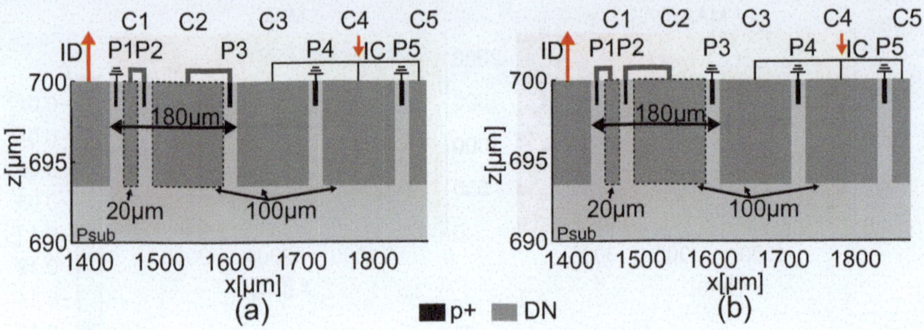

Fig. 7.17 Configuration of a double MAAP (**a**) and double NFA (**b**) active protection

Figure 7.18 shows the simulated characteristics of single and double structures. As also reported in [8], the use of a double MAAP does not improve the efficiency of the protection when compared to the single MAAP. The current injected into the substrate is almost the same in both cases (Fig. 7.18a).

Unlike with the single MAAP structure, the typical *gain kink* is not prominent in the lateral gain of the double MAAP (see Fig. 7.18b). Because of the double MAAP structure, the saturation of each single MAAP occurs for different current values resulting in two less marked kinks. The double MAAP gain is higher at the first kink, but it is lower at the second kink when compared to the single MAAP structure. This flattens out the overall lateral gain of the double MAAP.

The collected current over the reverse current is shown in Fig. 7.18d. The relative error between collected currents with single and double structures is also shown. For low reverse currents, the double NFA structure is 25% more efficient than the single NFA, but the single NFA is more efficient at high reverse currents, with a reduction of collected currents of up to 50%. On the other hand, single and double MAAP structures show the opposite behavior. These results are consistent with those reported in [8].

These results show that for a given isolation area, one single or two consecutive structures behave differently for low or high reverse currents. However, since the global objective is the reduction of currents collected by low-voltage collectors, two consecutive NFA structures are suitable for such an achievement.

7.4.4 Combined MAAP and NFA Active Guard Rings

The series combination of a NFA with a MAAP or a passive protection structure is investigated in this section. The operating principles and the efficiency of NFA and MAAP follow different processes. The NFA reduces the injection of reverse current into the substrate, while the MAAP reduces the lateral propagation. Thus, as also discussed in [8], the series combination of a NFA with a MAAP can reduce the collected current even more. The simulated structure with consecutive NFA and

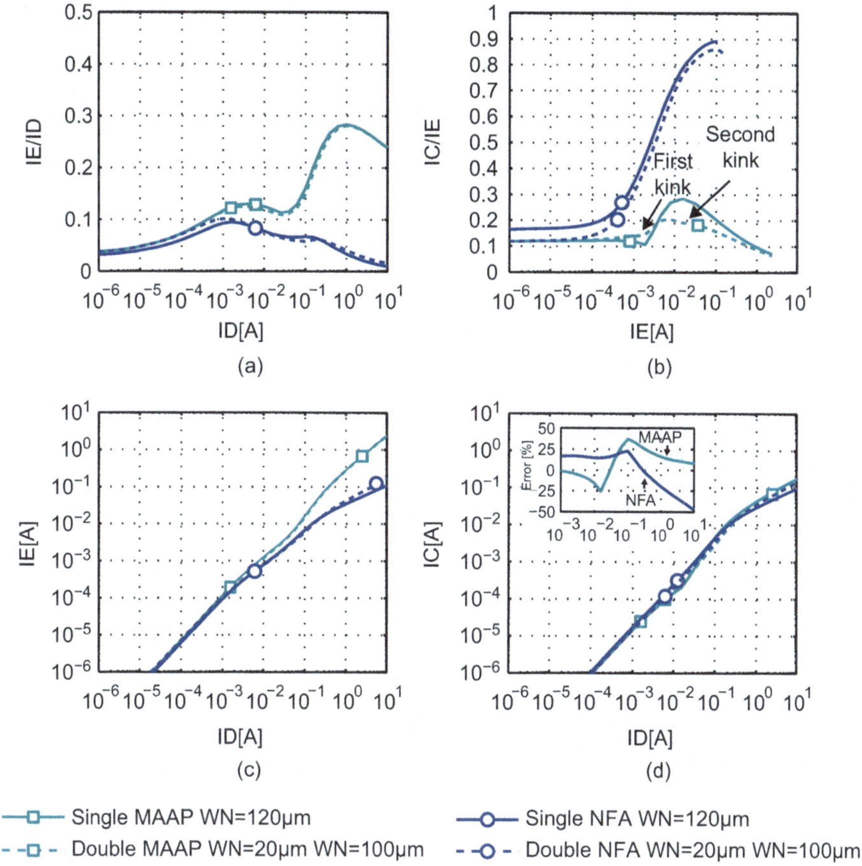

Fig. 7.18 Comparison of electrical characteristics of the double MAAP and double NFA active protections as a function of the reverse current I_D. Simulated injected current ratio (**a**), lateral NPN gain (**b**), total injected current (**c**), and total collected current (**d**)

MAAP is shown in Fig. 7.19a. The width of the NFA n-ring is 20 µm, while the width of the MAAP n-ring is 100 µm.

Furthermore, two additional structures with equal width are also considered for comparison and based on the Fig. 7.19b configuration.

- The first configuration combines a NFA 20 µm wide and a passive one 100 µm wide.
- The second configuration includes instead a NFA 20 µm wide, a passive n-ring 50 µm, and a MAAP 50 µm wide, which is explicitly shown in Fig. 7.19b.

Simulation results are shown in Fig. 7.20. The current injected into the substrate is comparable in all three cases (Fig. 7.20a, c). Nevertheless, the lateral gains plotted in Fig. 7.20b are different. The lateral gain of the structure NFA-MAAP follows the

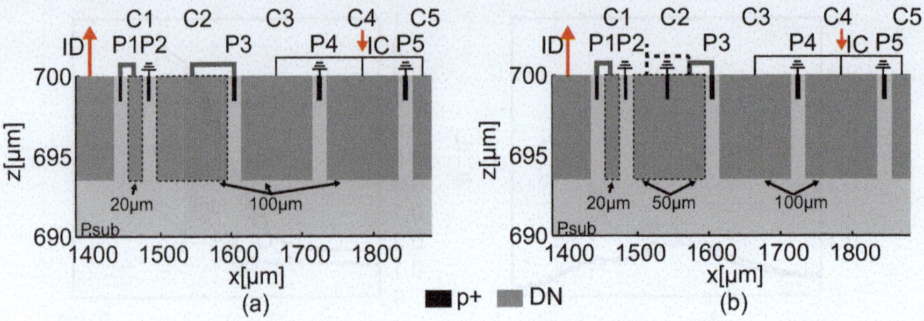

Fig. 7.19 Cross section of combined NFA-MAAP structures

same trend as the single NFA structure, and it is somewhat lower at higher currents (0.8 instead of 0.9). Indeed, the NFA structure pushes the current toward the MAAP protection which slightly lowers the gain since it saturates earlier.

If a passive protection comes after a NFA structure, then the lateral gain is significantly reduced. Moreover, if the 100 μm passive protection is further divided into a 50 μm passive ring and a 50 μm MAAP structure, then the lateral gain is further attenuated.

Individually these improvements might be small, but together they can add up to make a real difference. Now, if the total collected currents are compared to the reverse current I_D (see Fig. 7.20d), the solution with the combination of NFA, passive, and MAAP performs best. Unexpectedly, the solution with NFA and passive protection is the worst. This is due to the difference in the behavior of the level of injected current, which is higher when the reverse current increases.

With respect to the combination of NFA and passive protection, the relative error of remaining structures is plotted in Fig. 7.20d. The solution with the cascade of NFA, passive, and MAAP further reduces the collected current by 25% at high reverse currents. This comparison highlights the fact that the joint reduction of both factors results in a lower collected current.

Within structures of the same type, the vertical gain α_{IE} is higher than the lateral one α_{NPN}, or vice versa. Indeed, both passive and MAAP structures do not effectively reduce the injection of current into the substrate (high α_{IE}), but on the other hand, they limit the lateral propagation (low α_{NPN}). With NFA, due to the floating state of the substrate potential nearby the injection region, the effective injected current is lower (low α_{IE}). However, in this case nothing limits the lateral propagation of injected currents (high α_{NPN}).

Only then the combination of the two techniques does make it possible to lower both injection and propagation and thus improve the effectiveness of the protections. Obviously, many iterations are required for the optimization of the combined protecting structure according to the design needs.

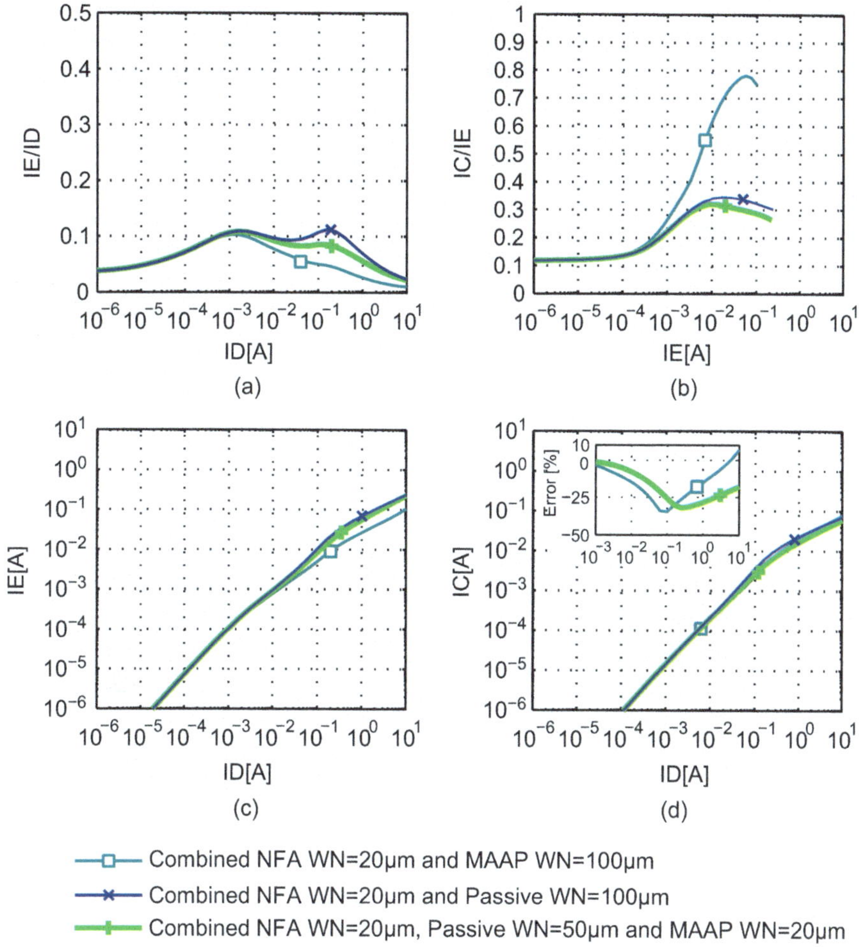

Fig. 7.20 Comparison of electrical characteristics of combined passive, MAAP, and NFA protections as a function of the reverse current I_D. Simulated injected current ratio (**a**), lateral NPN gain (**b**), total injected current (**c**), and total collected current (**d**)

7.4.5 *Protection Performance with Equal Width*

As discussed in Sect. 7.2, the simulation of the NDMOS with a n-well covering the entire chip provides initial information about the width of the passive n-ring needed to reduce the collected current to the required value. With equal width, the collected current can be further reduced by implementing a combination of passive and active protections. Therefore, for a direct comparison, the collected currents with different protection configurations are summarized on the same plot in Fig. 7.21. Among the simulations, those assuming a 180 μm wide region are considered.

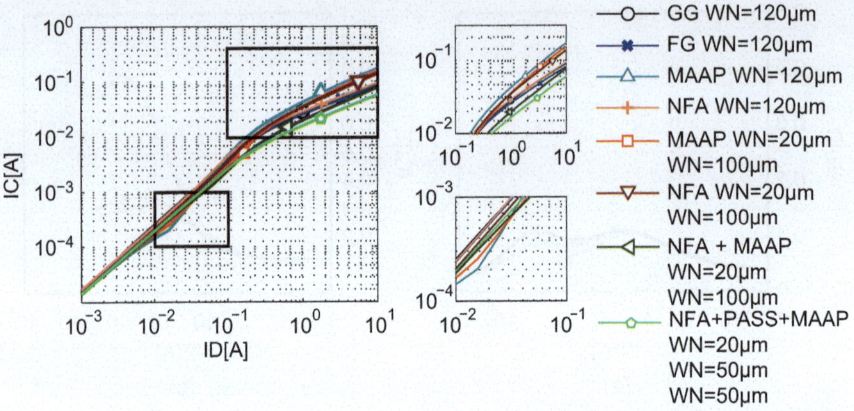

Fig. 7.21 Total collected current over NDMOS reverse current with a zoom at low and high reverse current

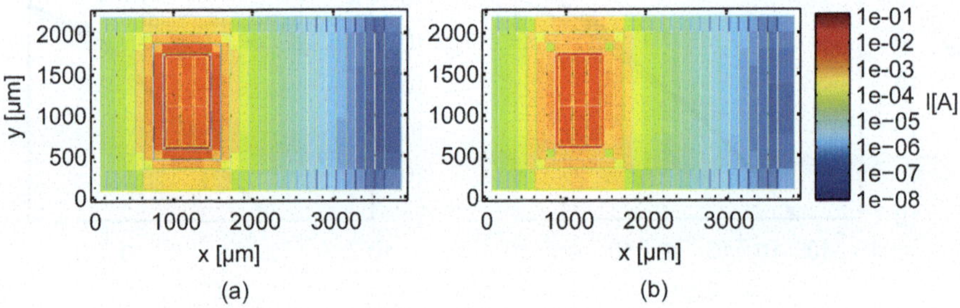

Fig. 7.22 Current distribution for a reverse current of 1 A with a standard passive protection 120 μm wide (**a**) and active NFA + MAAP protection 120 μm wide (**b**)

In this study, although most of collected current curves overlap, between different scenarios, the maximum variation at high currents amounts to 120 mA, whereas minimal variation occurs at low currents where it is negligible. It appears that at high reverse currents, the change in collected current when compared to various protection strategies can be up to 200%. The distribution of collected currents for a reverse current of $I_D = 1$ A is plotted in Fig. 7.22a, b. In the first layout, a single passive n-ring 120 μm wide encloses the NDMOS. In this case, the injected current is initially higher (higher α), and then it is clearly visible in the collection area at the edge of the NDMOS. When using a NFA structure followed by a MAAP, the injected current is lower, and in this case, the current is collected near the NDMOS as well but within a slightly larger area. The design of the optimal protection might require many iterations. Indeed, comparisons have to be made over a broader range of scenarios to draw general conclusions. However, the design process is fast since only the NDMOS layout with the selected ring has to be extracted for simulations of the substrate model.

7.5 Conclusions

In this chapter, a versatile approach for monitoring substrate currents and couplings in HV ICs with circuit-level simulations was presented. With this approach, parasitic substrate couplings are already evaluated during the circuit feasibility study, enabling IC designers to access valuable results in the early stage of IC design, where before such results could only be obtained in the final chip verification step. Furthermore, this analysis allows for the optimization of a circuit layout with minimal parasitic effects and thus for the estimation of passive and active protection effectiveness. A comprehensive investigation was conducted to evaluate the performance of the most used passive and active protections against reverse current injected during a NDMOS belowground condition. For this analysis, a number of important parameters were considered such as the width of the protections, the number of consecutive protections, as well as combined structures. Since protections against substrate couplings have to be optimized according to the design specifications and to the selected HV technology, generally valid rules and guidelines do not exist. This approach opens new horizons to design efficient and optimal protections and minimize couplings in HV ICs.

References

1. W. Horn, H. Zitta, A robust smart power bandgap reference circuit for use in an automotive environment. IEEE J. Solid State Circuits **37**(7), 949–952 (2002)
2. M. Schrems, M. Knaipp, H. Enichlmair, V. Vescoli, R. Minixhofer, E. Seebacher, F. Leisenberger, E. Wachmann, G. Schatzberger, H. Gensinger, Scalable high voltage CMOS technology for smart power and sensor applications. e & i Elektrotechnik und Informationstechnik **125**(4), 109–117 (2008)
3. R.J. Widlar, Controlling substrate currents in junction-isolated ICs. IEEE J. Solid State Circuits **26**(8), 1090–1097 (1991)
4. A. Hastings, *The Art of Analog Layout* (Prentice Hall, Lebanon, 2005)
5. International Organization for Standardization, ISO 7637-2 Road vehicles electrical disturbances from conduction and coupling Part 2: electrical transient conduction along supply lines only (2011)
6. S.H. Voldman, *Latchup* (Wiley, New York, 2008)
7. O. Gonnard, G. Charitat, P. Lance, M. Suquet, M. Bafleur, J.-P. Laine, A. Peyre-Lavigne, Multi-ring active analogic protection for minority carrier injection suppression in smart power technology, in *Proceedings of the 13th International Symposium on Power Semiconductor Devices and ICs* (2001), pp. 351–354
8. T.K.H. Starke, P.M. Holland, S. Hussain, W.M. Jamal, P.A. Mawby, P.M. Igic, Highly effective junction isolation structures for PICs based on standard CMOS process. IEEE Trans. Electron Devices **51**(7), 1178–1184 (2004)
9. J. Wittmann, C. Rindfleisch, B. Wicht, Substrate coupling in fast-switching integrated power stages, in *IEEE 27th International Symposium on Power Semiconductor Devices & IC's (ISPSD)* (IEEE, New York, 2015), pp. 341–344
10. S.H. Voldman, C.N. Perez, A. Watson, Guard rings: theory, experimental quantification and design, in *Electrical Overstress/Electrostatic Discharge Symposium (EOS/ESD)*, Sept. 2005 (IEEE, New York, 2005), pp. 1–10

11. B. Murari, F. Bertotti, G.A. Vignola, *Smart Power ICs: Technologies and Applications*, vol. 6 (Springer Science & Business Media, Berlin, 2002)
12. O. Gonnard, G. Charitat, P. Lance, E. Stefanov, M. Suquet, M. Bafleur, N. Mauran, A. Peyre-Lavigne, Substrate current protection in smart power IC's, in *Proceedings of the IEEE 12th International Symposium on Power Semiconductor Devices and ICs (ISPSD)* (2000), pp. 169–172
13. G. Charitat, Isolation issues in smart power integrated circuits, in *23rd International Conference on Microelectronics*, vol. 1 (2002), pp. 15–22
14. S. Gupta, J.C. Beckman, S.L. Kosier, Improved latch-up immunity in junction-isolated smart power ICs with unbiased guard ring. IEEE Electron Device Lett. **22**(12), 600–602 (2001)

Appendix A
Substrate Model Parameters

The generalized lumped device parameters are dependent on the technology process choices (like the doping profiles, the well depths, etc.) and on the type of semiconductor (e.g., mobility is different for silicon and germanium). In the following the complete list of parameters and numerical constants (Table A.1) used is reported.

A.1 Intrinsic Parameters for Silicon

The EPFL substrate model, developed for silicon technologies, requires the intrinsic parameters of silicon, namely, the intrinsic carrier concentration, the mobilities for electrons and holes, and the corresponding lifetimes. As a matter of fact, these quantities are strongly doping and temperature dependent. Given the temperature T in [K] and the doping concentration N in [cm^{-3}] as inputs, the following empirical relations from literature have been considered:

- **Normalized temperature [–] and Thermal voltage [V]**

$$T_n = \frac{T}{T_{\mathrm{ref}}} \quad V_t = \frac{kT}{q} \tag{A.1}$$

- **Energy gap [eV]** [1]

$$E_g = 1.1696 - 0.000473\frac{T^2}{T+636} - \Delta E_g \tag{A.2}$$

- **Bandgap narrowing [eV]** [2] (only for $N > 3.162 \times 10^{18}$ cm^{-3})

$$\Delta E_g = 0.00684\ln^2\frac{N}{3.162 \times 10^{18}} \tag{A.3}$$

P. Buccella et al., *Parasitic Substrate Coupling in High Voltage Integrated Circuits*, Analog Circuits and Signal Processing, https://doi.org/10.1007/978-3-319-74382-0

Table A.1 Basic physical constants

Parameter	Symbol	Value	Unit
Permittivity of silicon	ε_s	1.0359×10^{-10}	F/m
Elementary charge	q	1.6022×10^{-19}	C
Boltzmann constant	k_B	1.3807×10^{-23}	J/K
Reference temperature	T_{ref}	300	K

- **Intrinsic carrier concentration [cm^{-3}]** [3]

$$n_i = 1.075 \times 10^{10}\, T_n^{1.5} \exp\left(\frac{E_g(T_{\text{ref}})}{2V_t(T_{\text{ref}})} - \frac{E_g}{2V_t}\right) \tag{A.4}$$

- **Mobility (electrons and holes) [cm^2/V s]** [4]

$$\mu_n = 88.0\, T_n^{-0.57} + \frac{1252\, T_n^{-2.33}}{1 + [N/(1.26 \times 10^{17}\, T_n^{2.4})]^{0.88\, T_n^{-0.146}}} \tag{A.5a}$$

$$\mu_p = 54.3\, T_n^{-0.57} + \frac{407\, T_n^{-2.23}}{1 + [N/(2.35 \times 10^{17}\, T_n^{2.4})]^{0.88\, T_n^{-0.146}}} \tag{A.5b}$$

- **Lifetime SRH (electrons and holes) [s]** [5]

$$\tau_{n,\text{SRH}} = \frac{30 \times 10^{-6}}{1 + N/10^{17}} T_n^{1.5} \quad \tau_{p,\text{SRH}} = \frac{10 \times 10^{-6}}{1 + N/10^{17}} T_n^{-1.5} \tag{A.6}$$

All the relevant temperature and doping dependencies are covered with these expressions. As a consequence no additional temperature fitting coefficients will be added in the model. Nonetheless, it is necessary to make a comment regarding the carrier lifetime since different fitting values are found in the literature. The general Scharfetter equation (i.e., empirical relations between carrier lifetime and doping concentration) is given in Eq. (A.7). It is well known that the higher the doping, the lower is the lifetime until it reaches the Auger recombination limit for very high doping concentrations. Moreover the holes' lifetime is always lower than the electrons' lifetime by about a factor 3. However, the maximum lifetime is strongly dependent on the process, and discordant values are found in published works. In power devices maximum carrier lifetimes τ_0 of the order of microseconds are registered compared to millisecond lifetimes measured for optical generated processes [6]. In conclusion this parameter should be definitively left as a calibration parameter.

$$\tau_m = \frac{\tau_0}{1 + N/10^{17}} T_n^{\alpha} \tag{A.7}$$

Also for the lifetime temperature dependency, no general models can be used. Several works [7, 8] showed a positive exponent α that should be properly fitted for a given technology. This means that for high temperatures lifetime increases, and the same occurs for the minority carrier couplings.

A.2 Instance and Model Parameters

In each lumped component, there are technology parameters (to be calibrated with respect to different technologies, as discussed in Chap. 5) and geometrical parameters which depend on layout and process cross section. These geometrical parameters (namely, the area and the length of each component) should be computed by the substrate extraction tool since they are strongly dependent on the substrate meshing strategy. In the following tables, the meshing size is assumed to be between 100 nm and 500 μm. The technology parameter ranges are listed with typical values for HV technologies. The resistor and homojunction lumped models can be used for both N- or P-doped semiconductors with a *type* parameter which is −1 for N-doped and +1 for P-doped silicon. Notice that the only AC parameters are V_{bi}, m, x_0 used in the diode which are equivalent to the ones used in standard SPICE models for the junction capacitance C_j [9] (Tables A.2, A.3, and A.4).

Table A.2 EPFL resistance parameters

	Symbol	Unit	Min.	Typ.	Max.
Instance parameters					
Resistor area	A	m^2	1e−14	1e−10	2.5e−8
Resistor length	L_r	m	100n	10u	500u
Geometrical ratio	R_g	–	2e−4	1	5e3
Dimensional parameter (pin1)	dm_1	–	2	6	3
Dimensional parameter (pin2)	dm_2	–	2	6	3
Model parameters					
Doping	N	cm^{-3}	1e14	1e15	1e21
Minority carrier lifetime	τ_0	s	1n	10u	1m
Type (−1, N-doped; 1, P-doped)	*type*	–	−1	–	+1

Table A.3 EPFL diode parameters

	Symbol	Unit	Min.	Typ.	Max.
Instance parameters					
Junction area	A	m^2	1e−14	1e−10	2.5e−8
P-side length	W_p	m	100n	10u	500u
N-side length	W_n	m	100n	5u	500u
Dimensional parameter (P-side)	dm_p	–	2	6	3
Dimensional parameter (N-side)	dm_n	–	2	6	3
Model parameters					
Doping (P-side)	N_a	cm^{-3}	1e14	1e15	1e21
Doping (N-side)	N_d	cm^{-3}	1e15	1e17	1e21
Recombination lifetime	τ_r	s	1n	10u	1m
Minority carrier lifetime	τ_0	s	1n	10u	1m
Built-in potential	V_{bi}	V	0.4	0.7	1.4
Junction doping grading	m_j	–	1.5	2	3.5
AC junction parameter	x_0	m	1n	1u	100u
SNS current parameter	n	–	1	0.5	3
Breakdown voltage	BV	V	3	10	100

Table A.4 EPFL homojunction parameters

	Symbol	Unit	Min.	Typ.	Max.
Instance parameters					
Resistor area	A	m^2	1e−14	1e−10	2.5e−8
Low-side length	L_1	m	100n	10u	500u
High-side length	L_2	m	100n	200n	500u
Dimensional parameter (pin1)	dm_1	–	2	6	3
Dimensional parameter (pin2)	dm_2	–	2	6	3
Model parameters					
Doping (low-side)	N_1	cm^{-3}	1e14	1e15	1e21
Doping (high-side)	N_2	cm^{-3}	1e14	1e19	1e21
Minority carrier lifetime	τ_0	s	1n	10u	1m
Type (−1, N-doped; 1, P-doped)	*type*	–	−1	–	+1

References

1. C.D. Thurmond, The standard thermodynamic functions for the formation of electrons and holes in Ge, Si, GaAs , and GaP. J. Electrochem. Soc. **122**(8), 1133–1141 (1975)
2. H.S. Bennett, C.L. Wilson, Statistical comparisons of data on band-gap narrowing in heavily doped silicon: electrical and optical measurements. J. Appl. Phys. **55**(10), 3582–3587 (1984)

3. M.A. Green, Intrinsic concentration, effective densities of states, and effective mass in silicon. J. Appl. Phys. **67**(6), 2944–2954 (1990)
4. N.D. Arora, J.R. Hauser, D.J. Roulston, Electron and hole mobilities in silicon as a function of concentration and temperature. IEEE Trans. Electron Devices **29**(2), 292–295 (1982)
5. M.E. Law, E. Solley, M. Liang, D.E. Burk, Self-consistent model of minority-carrier lifetime, diffusion length, and mobility. IEEE Electron Device Lett. **12**(8), 401–403 (1991)
6. M. Schenkel, Substrate current effects in smart power ICs. PhD thesis, ETH Zürich, Nr. 14925, 2003
7. J. Lutz, H. Schlangenotto, U. Scheuermann, R. De Doncker, *Semiconductor Power Devices: Physics, Characteristics, Reliability* (Springer, Berlin, 2011)
8. W. Zimmermann, Temperature dependence of carrier lifetime in silicon power devices. Phys. Status Solidi A **10**(1), K49–K51 (1972)
9. C. Salamero, N. Nolhier, A. Gendron, M. Bafleur, P. Besse, M. Zecri, TCAD methodology for ESD robustness prediction of smart power ESD devices. IEEE Trans. Device Mater. Reliab. **6**(3), 399–407 (2006)

Index

P. Buccella et al., *Parasitic Substrate Coupling in High Voltage Integrated Circuits*, Analog Circuits and Signal Processing, https://doi.org/10.1007/978-3-319-74382-0

The manufacturer's authorised representative in the EU is Springer Nature Customer Service Centre GmbH, Europaplatz 3, 69115 Heidelberg, Germany. If you have any concerns regarding our products, please contact ProductSafety@springernature.com

Printed and bound by CPI Group (UK) Ltd, Croydon, CR0 4YY
29/07/2026
02180578-0001